TWO CENTURIES
OF FEDERAL INFORMATION

**PUBLICATIONS IN
THE INFORMATION SCIENCES**

Rita G. Lerner, Consulting Editor

TWO CENTURIES OF FEDERAL INFORMATION/*Burton W. Adkinson*
INDUSTRIAL INFORMATION SYSTEMS: A Manual for Higher Managements
 and Their Information Officer/Librarian Associates/*Eugene B. Jackson and
 Ruth L. Jackson*
LIBRARY CONSERVATION: Preservation in Perspective/*John B. Baker and
 Marguerite C. Soroka*

TWO CENTURIES OF FEDERAL INFORMATION

Burton W. Adkinson

Q224.3
U6
A35
1978

Dowden, Hutchinson & Ross, Inc.
Stroudsburg, Pennsylvania

Copyright © 1978 by Dowden, Hutchinson & Ross, Inc.
Library of Congress Catalog Card Number: 78-7294
ISBN: 0-87933-269-7

All rights reserved. No part of this book may be reproduced or transmitted in any form or by any means—graphic, electronic, or mechanical, including photocopying, recording, taping, or information storage and retrieval systems—without written permission of the publisher.

80   79   78   1   2   3   4   5
Manufactured in the United States of America.

Library of Congress Cataloging in Publication Data

Adkinson, Burton W
    Two centuries of Federal information.
    Bibliography: p.
    1. Science—Information services—United States—History. 2. Technology—Information services—United States—History. 3. Science and state—United States—History. 4. Technology and state—United States—History. I. Title. II. Title: Federal information.
Q224.3.U6A35   353.008'1   78-7294
ISBN 0-87933-269-7

Distributed world wide by Academic Press,
a subsidiary of Harcourt Brace Jovanovich,
Publishers.

Preface

Many persons have asked why I chose to emphasize the federal government's scientific and technical (sci-tech) information policies and programs during the thirty-two-year period, 1942-1972, in this volume. There are several reasons for this choice. First, I am most familiar with this facet of the information industry and was a participant during this entire era. Second, the federal government's policy and program decisions during this period fostered the development of the most powerful and sophisticated sci-tech information services in the world. Third, it seemed useful to describe the symbiotic relationship among the several sectors of the information industry. Tensions created by competition among two or more sectors have frequently obscured the cooperative and mutual assistance activities among the for-profit organizations, the scientific and professional societies, the state and local governments, the universities, and the private foundations. However, no one of the sector could have accomplished the information systems development that was achieved cooperatively. More importantly, federal funds and federal personnel's initiative and ideas played a major role in encouraging every sector to contribute to the growth, advancement, and quality of present day sci-tech information services and products.

I wish to stress that during this period, sci-tech information programs in the nonfederal sector developed in a dramatic fashion. Scientific and professional societies, private libraries, and information services in private organizations experienced an almost explosive growth. Also, as in the case of federal services, some of these became the world's leading information services in their scientific or technical field. The nonfederal services had an important impact on U.S. government sci-tech information programs and policies. Although the development and contributions of these nonfederal services are equally as interesting and significant as those in the federal government, little of this nonfederal story is included in this book. It is my hope that some other author will prepare a chronicle of these important nonfederal contributions. Another facet of this thirty-two-year period that warrants review is the spectacular evolution of information processing technology. Although numerous articles and books have been published on these technical aspects, no one has yet prepared an overview of the impact of these technical innovations on the information industry. I hope a competent specialist will elect to do this complicated and difficult task.

This author gratefully acknowledges the friendly and able assistance received from persons both within and outside the federal government, without whose aid this book could not have been prepared. Several agencies' internal records were made available as were the private files of individuals who also searched their memories to assist in identifying details related to specific programs or policies. It is not feasible to name all of these persons, but the following warrant special recognition: Andrew Aines, Edward Brady, Helen Brownson, Douglas Berninger, Lee Burchinal, Mary Corning, Joseph Caponio, Martin Cummings, Melvin Day, Bernard Fry, John Davis, Margaret Fox, John C. Green, Madeleine Henderson, Bert Huntoon, David Hersey, Eugene Jackson, William Knox, David Lide, Bert Mulcahy, Eugene Miller, Marvin McFarland, Leila Moran, Foster Mohrhardt, John Price, Eugene Pronko, Robert Shannon, Hubert Sauter, Robert B. Stegmaier, Jr., R. A. Spencer, Gerald Sophar, Edward Stokely, Howard Twilley, Peter Urbach, and Harold Wooster. All of these persons are or were involved in federal information programs and willingly furnished information that would not otherwise have been available.

Special thanks are extended to the Library of Congress whose staff was unusually cooperative. Marvin McFarland and John Price were particularly helpful in locating material, arranging for working space, and introducing me to other staff members who assisted me. Also the staff of the Smithsonian Science Information Exchange and Robert Shannon of Oak Ridge Extension of the Atomic Energy Commission went to considerable trouble to locate and make available documents from files that proved essential in preparation of dicussions on the development of these agencies. The following persons outside the federal government were also particularly helpful in providing information: Dale Baker, Lowell Hattery, Edward Kennedy, John Creps, William Koch, Phyllis Parkins, William Steere, and Judith Werdell.

As the reader can surmise, the sources of information used in preparation of this book are numerous and varied. This variety of sources presented a problem in citations in this work. Since each topic is based on numerous sources, and I did not wish to interrupt the text with many references, I cited only those sources that proved particularly helpful when preparing the final version of the text. Several hundred additional references were consulted during preparation. A few of the published sources are listed in the supplementary material at the end of the text.

Rita Lerner, Andrew Aines, and Madeleine Henderson took time to read and comment on the completed manuscript. Their assistance was of great value as was that of Dwight E. Gray and Karen Reixach who spent many hours reading, editing, and commenting on parts of the manuscript as it was in preparaton and then reviewed the completed version. Their contributions significantly aided in the improvement of the content and presentation. However, I am responsible for the selection, organization, and presentation. Any errors of commission or omission are mine. Finally, if it were not for the persistent and friendly persuasion of Rita Lerner, this book probably would not have been written.

In conclusion I wish to emphasize that individuals plan and execute programs and policies. I consider it an honor to have been a part of an exciting federal program as it was rapidly developing. In found most people who participated in this era of sci-tech information activities to be dedicated to the advancement of information services. I hope the presentation in this book does justice to the contributions of the hundreds of participants. It is my further hope that this book will stimulate others to prepare publications on other facets of this large, complex, and rapidly evolving sector of human activities.

Burton W. Adkinson

Contents

Preface v

Chapter 1 Introduction 1
Chapter 2 The Beginnings: Federal Information Services, 1790-1942 9
    Patent Office 11
    Library of Congress 12
    Smithsonian Institution 16
    Coast and Geodetic Survey 20
    Geological Survey 21
    Early Military Projects 23
    Department of Agriculture 24
    Conclusion 27
    Notes and References 27
Chapter 3 Federal Sci-Tech Information Centers, 1942-1956 29
    Defense Documentation Center 30
    National Technical Information Service 33
    Medical Sciences Information Exchange 37
    Biological Sciences Information Exchange 39
    Library of Congress 40
    Technical Information Division, Atomic Energy Commission 42
    Other Federal Agencies 48
    Conclusion 51
    Notes and References 51
Chapter 4 The Decade of Investigation, Reorganization, and Expansion, 1957-1966 53
    Investigations at the National Level 54
    Policy and Organizational Changes 61
    Operational Changes: Introduction of New Technology 66
    R&D Projects and Studies 72
    Conclusion 75
    Notes and References 75
Chapter 5 Consolidation, Computerization, and Retrenchment, 1966-1972 79
    Program and Policy Changes 79
    Operational and Technical Advances 85

x Two Centuries of Federal Information

Reduction of R&D Support for Library and Information Sciences	88
Conclusion	90
Notes and References	91
Chapter 6 Federal R&D Programs and Studies	93
Patent Office	96
Report of the Baker Panel of the President's Science Advisory Committee	100
The Crawford Task Force Report and the Role of the Office of Science and Technology	101
Weinberg Report to the President's Science Advisory Committee	102
The Fry and Heller Studies	103
The National Systems Task Force	107
UNISIST	107
Conclusion	110
Notes and References	111
Chapter 7 Relationship between Federal Agencies and Nonfederal Organizations	113
The Expansion of Federal Agencies' Sci-Tech Information Programs	114
Advisory Panels and Study Groups	118
Systems, Program Development, and Operational Experiments	119
Federal Assistance to Organizations in Sci-Tech Information Activities	121
National Academy of Sciences	125
Notes and References	130
Chapter 8 The Federal Government and Foreign Exchange of Information	131
Cooperative Exchange Projects with Foreign Organizations	132
The Federal Government and Nongovernmental Organizations	136
Sci-Tech Information Programs of Other Countries	137
International Governmental Organizations	143
Conclusion	144
Notes and References	145
Chapter 9 Trends, Persons, and Future Directions	147
Trends	148
Federal Personnel Influential in Sci-Tech Information	155
Possible Future Developments	162
Bibliography	165
Appendix A Chronology	171
Appendix B Acronyms and Abbreviations	181
Appendix C White House Press Release Containing Baker Report	185

Appendix D	Excerpts from the Crawford Report	197
Appendix E	Excerpts from the Weinberg Report	199
Appendix F	Excerpts from the SATCOM Report	207
Appendix G	Excerpts from the Kennedy Report	221

Index 225

TWO CENTURIES
OF FEDERAL INFORMATION

Chapter 1

Introduction

One of the most important products of exploration and research and development (R&D) is information. Man, using this information, has increased his knowledge and understanding of the earth and its wealth of resources. Not content with his investigation of the immediate environment, man now explores the solar system in trying to understand its origin and characteristics. One of the prime purposes is to identify the relationships among these phenomena and their impact on his earthly environment. This drive for information and knowledge has stimulated man to develop technology that enables him to explore more deeply into the micro- and macrocosms surrounding him.

Although individual people are responsible for the development of hypotheses, concepts, and theories that increase man's ability to understand and use resources, team effort is usually required for exploration and R&D. Thus, governments became involved in these efforts early and soon established mechanisms not only to perform the exploration and the R&D but to house, preserve, organize, and disseminate the accumulated information. Recognition of the value of exploration and R&D and the principle results of R&D—scientific and technical (sci-tech) information—stimulated national governments to sponsor and support learned societies, national libraries, archives, and museums.

Among the well-know foreign scientific institutions are the British Royal Society, the French Academy of Sciences, the German Max Planck Institute, the USSR Academy of Sciences, the Science Council of Japan, and the National Research Council of Canada. These organizations each have different relationships and fulfill different roles for their respective governments. For example, the British Royal Society, the French National Academy of Sciences, and the Science Council of Japan, in addition to being prestigious learned societies, are also used by their governments for advice on sci-tech programs. The USSR Academy of Sciences, the Max Planck Institute, and the Canadian National Research Council have been responsible for operating laboratories and maintaining sci-tech libraries and other sci-tech information services. In recognition of the need to evaluate, organize, and disseminate the results of sci-tech R&D, each of the above institutions issue high-quality journals and monographs.

Most national governments maintain national libraries, museums, and other information resources and services that serve their own needs and frequently are used by citizens of other countries. For example, the United Kingdom has the newly organized British Library that includes the Library of the British Museum and the National Lending Library; France maintains the Biblioteque Nationale and the information activities of the National Center for Scientific Research (CNRS); the USSR has among its information services the famous Lenin Library and the information activities of the Academy of Sciences, including the well-known All Union Institute of Scientific Information (VINITI); the Canadian National Research Council has been a sponsor of a National Science Library and a Technical Information Service; and Japan maintains the National Diet Library and the Science and Technical Agency with its Japan Information Center for Science and Technology (JICST).

Since World War II, most large, technically advanced countries, as well as many smaller countries, have developed mechanisms to coordinate and frequently to plan and develop national sci-tech information programs. The USSR has its department of Scientific and Technical Information and Propaganda as a unit of the State Committee for Science and Technology. Until 1965 the United Kingdom relied on its Department of Scientific and Industrial Research (DSIR) to plan and develop its government-sponsored sci-tech information activities. In that year the Office of Scientific and Technical Information (OSTI) was established in the Ministry of Education and Science to strengthen planning and coordination of the United Kingdom's sci-tech information activities. To improve national coordination further, in 1972 the United Kingdom organized the British Library in which OSTI became a department of R&D. The Canadian government designated its NRC library as the National Science Library during the postwar period and since 1965 has mounted a number of studies and taken actions to improve operations and coordination among its sci-tech information services. Many other national governments have moved to improve their planning and operations of sci-tech information services, and France, Germany, the Netherlands, Japan, and Australia have all been involved in similar programs.

While the United States has also taken many actions to improve its sci-tech and other information activities since World War II, the U. S. approach has been unique because of the diversity of organizations that have played important roles in the movement. Scientific and professional societies; commercial and industrial organizations; educational institutions; private research establishments; and metropolitan, state, and national governments all have been involved in the efforts to strengthen and coordinate scientific information resources and services. The purpose of this book is to turn the spotlight on the federal government's role during this period.

The major focus of this book is federal policies and programs in the scientific information field. Decisions are identified; where the information is available, important factors leading to these decisions are discussed; and the results of these decisions are described. For example, the Library of Congress was destroyed when the Capitol Building was burned during the War of 1812. A debate immediately ensued in the Congress as to whether a small congressional reference library should be organized or whether Thomas Jefferson's offer to sell his library to Congress should be accepted. Jefferson's library was one of the more important personal libraries in the United States and material therein reflected his broad interests with an emphasis on science. The result of the debate was the purchase of the library and the acceptance of the concept that the congressional library should be more comprehensive in scope than merely a small working collection organized to meet the immediate needs of Congress. This policy decision has been adhered to in later expansions of the Library of Congress's collections and services. Another series of debates occurred in Congress over the role of the Office of Weights and Measures in the Treasury Department. Many members of Congress wished to limit the activities of this office to establishing standards of weights and measures for custom administration; other congressmen argued that the office should be the national center for determining standards used in the United States as a whole. History indicates that the broader concept was accepted—a decision that led ultimately to establishment of the National Bureau of Standards.

Detailed descriptions of the development of federal libraries or other scientific or technical information activities are not included in this book. Instead, a few activities have been selected to illustrate the policy and program decisions that led to the growth of some of the larger federal sci-tech information services. Review of all federal sci-tech libraries and other similar services would have been too voluminous as well as unnecessary to illustrate the impact of policy and program decisions. Decisions affecting nonscientific federal library development are not discussed unless they are considered important to an explanation and analysis of sci-tech information activities.

Because scientific library and other information services in the government were initiated primarily to further major programs of federal agencies, there was little coordination and much duplication throughout their development. During and immediately following World War II, several large technical information centers were organized in the three branches of the military services as well as in the Department of Commerce and the Atomic Energy Commission. Some of these duplicative activities were consolidated and coordinated between 1950 and 1972. However, neither the congressional nor executive branches of the government developed a uniform federal sci-tech information program, nor was any serious attempt made to establish a closely coordinated federal sci-tech research program. The thinking seemed to be that several agencies working on the same problem would expedite the solution. Since the information and library centers were organized to serve R&D, they exhibited the same uncoordinated characteristics as those of the parent programs.

Although Congress frequently included library and information services in its authorization of new agencies, it often was very reluctant to fund such activities adequately. An example is the passage by Congress in 1950 of PL-776 (81st Cong., 2nd sess. S.64; Stat. 823), which gave the Department of Commerce broad responsibilities for collecting and disseminating sci-tech material. This law was the legislative base upon which the Office of Technical Services (OTS) operated. Yet Congress never appropriated adequate funds to implement even the minimum requirements of this legislative mandate.

Within this book, certain areas of the sci-tech information field are discussed very briefly or have been excluded altogether. One exclusion is the topic of personal and informal information transfer, including attendance at and support of scientific and technical meetings, as well as small group discussions by scientists and engineers. In appropriations acts, Congress from time to time has restricted use of funds for such purposes, but neither it nor the executive branch agencies have stated or followed a consistent policy in this regard. Yet in terms of time alloted and of direct and indirect costs, attendance at meetings and informal meetings probably rank near the top of federal expenditures for sci-tech information activities. However, the author has been unable to find adequate data to warrant a comprehensive discussion of this area.

An area that is considered in this book more briefly than its importance would warrant concerns sci-tech libraries. Some attention is given to the federal policies related to these libraries. However, since another author is preparing a study on federal libraries, an extended discussion of this subject is not offered here, even though libraries are a vital part of the total federal sci-tech information complex.

Another area that is only partially considered in this book is that of numerical data collection, analysis, and dissemination. This topic is complicated and many of its activities are an integral part of R&D. While little attention is given to the gathering, storing, and analyzing of unevaluated data herein, programs and policies related to the organization of and services for evaluated (critical) data are covered.

The problem of organizing the information in this volume proved exceedingly complex. After studying various alternative methods, the author decided to use an historical approach in Chapters 2 through 5, which discuss the establishment of large federal libraries and information centers. A topical presentation is employed for Chapters 6 through 9, which include discussion of such subjects as R&D in information sciences and the federal government's interactions with U. S. and foreign organizations and finally conclude with a discussion of future trends.

For the reader's convenience, the first group of chapters on the development of federal policies and programs is divided into somewhat arbitrary periods. These intervals are not based on clear lines of demarcation of federal sci-tech information activities, but rather on the emphasis given to various aspects of development of libraries and information centers during each period. Consequently there is some overlap in activities from one period to another.

Chapter 2 presents a brief survey of library and information activities from 1790 to 1942 (approximately the beginning of World War II). During this interval most major governmental libraries were established. Among these were the Library of Congress, the Department of Agriculture Library, the Commerce Department Library, the Army Surgeon General's Library (later the National Library of Medicine) and specialized libraries in the Department of Interior (e.g., those of the Geological Survey and the Bureau of Mines). Also many of the major agencies that gather, organize, and disseminate information were developed. These include the Patent Office, the Copyright Office, the Coast and Geodetic Survey, the Weather Bureau, the Geologic Survey, the Bureau of Standards, and the National Advisory Committee for Aeronautics (later the National Aeronautics and Space Administration.)

The second interval, discussed in Chapter 3, covers sci-tech information developments from World War II to 1957 when Sputnik was orbited. This period was characterized by establishment of the forerunners of the large sci-tech centers, such as the present National Technical Information Service, the Defense Documentation Center, the Science and Technology Division in the Library of Congress, the Atomic Energy Commission's Technical Information Division, the Office of Science Information Service of the National Science Foundation, and the predecessor agencies of the Smithsonian Science Information Exchange. The production and wide use of the "technical report" was an important phenomenon of this period. Libraries and information centers began operational use of the rapidly advancing photographic techniques and were experimenting with new electronic technologies.

Chapter 4 reviews the "Post-Sputnik Era," 1957 to 1966, during which large and varied support programs for R&D in library, information, and communications sciences developed. Congressional committees and the Executive Office of the President exerted pressure on agencies for more orderly and better coordinated programs among sci-tech information services within or fostered by the federal government. This period was characterized by many congressional and executive branch studies and investigations into sci-tech information activities. At this time many private information-oriented companies were founded that were used by federal agencies for management and systems studies and for specialized information services. Decisions were made to expand the role of the National Library of Medicine, to enlarge that of the National Science Foundation, to establish the sci-tech information program of the National Aeronautic and Space Administration, and to allow surplus foreign credits to be used for translations and library acquisitions. Federal information agencies began exploratory cooperative programs with their counterparts in European countries and began to use international organizations—such as the Organization for Economic Cooperation and Development, Unesco, EURATOM, Food and Agriculture Organization, and International Council of Scientific Unions—as forums for discussion of common problems and as catalysts for development of multinational information systems. Coordination of federal information activities increased through the appointment of a special assistant for information activities in the Office of the Science Advisor to the President and through the formation of an interdepartmental committee to stimulate planning of cooperative information programs. During this period, expectations grew for rapid development of information systems based on new photographic and electronic technologies—an attitude that led to the premature introduction of mechanized techniques into some operating information systems. In many cases these techniques had to be abandoned temporarily or their use markedly curtailed or modified.

The final historically oriented chapter (Chapter 5) reviews the period from 1966 to 1972, a span markedly different from the preceding one. First, library and technical information centers expanded their use of the new technologies—at first to improve efficiency in publication and later for actual information retrieval. These new systems paved the way for rapid and extensive exchange of bibliographic and numerical data among agencies. During this period, R&D funds first levelled off and then declined in several areas of scientific R&D, thereby leading to reappraisal of governmental science information programs. Increased attention was given to modifying or expanding information programs to meet such needs as information for managing and planning R&D programs so as to permit better coordination and cooperation among federal information systems, to conserve funds, and to develop services in support of environmental R&D programs. Federal support for nonfederal libraries and information services was given close scrutiny by the Bureau of the Budget and by various congressional committees.

Each of the next three chapters discusses one aspect of library and information sciences. Chapter 6 concentrates on R&D in this field. In spite of the absence of effective coordination and lack of clearly defined programs, the R&D results of this period laid the base upon which current information systems are founded. Therefore, they merit review. The chapter discusses the following topics: changes in R&D emphasis during this period and the prominent federal administrators of information sciences R&D. It also identifies the persons and organizations that were performing R&D, shows the impact of federally sponsored studies on the size and direction of R&D programs, and describes the efforts of the larger funding agencies to coordinate their support programs. One agency's information R&D programs is described to show how the political and funding problems influenced the direction and character of this program. The final topic in this chapter is a brief review of the initiation of Unesco's UNISIST program.

Chapter 7 examines the federal government's interaction in the library-information sciences with organizations in the private sector. First, is the federal government's dependence on scientific and professional societies for advice and information services; this relationship began when the government was first formed and still continues. Further, many of the for-profit, information-oriented companies have been called upon by federal agencies to perform a variety of tasks under contract. There are many instances when federal

Introduction 7

agencies assisted scientific and professional societies to improve their information activities or initiate new ones in order to further federal programs. The federal government turned either to scientific society or to commercial publishers as outlets for its R&D findings. Also discussed in Chapter 7 is the role of the National Academy of Sciences in its cooperative programs with federal agencies.

The third chapter in this group (Chapter 8) deals with various aspects of sci-tech information exchange between the United States and many foreign countries, some of which adopted U. S. techniques and systems to their own situations. In other cases there have been cooperative programs between a U. S. agency and its foreign counterpart. Conversely, the United States has adjusted its program as the result of information activities in such countries as the United Kingdom, the USSR, France, West Germany, and the Scandinavian countries. Increasingly, the United States has turned to international governmental and nongovernment organizations as forums for discussing common problems and as mechanisms for initiating international information services and systems. Examples are the development of the International Nuclear Information Service (INIS), Agri-Doc in Agriculture, CODATA in the critical data areas, and UNISIST—a world information network being sponsored by Unesco and ICSU.

The final chapter (Chapter 9) identifies trends in the library and information sciences and forecasts possible developments in this large and complex area.

Chapter 2

The Beginnings: Federal Information Services, 1790-1942

During the period between 1790 and 1942 the federal government, by congressional action or by executive or departmental order, established several hundred federal libraries or specialized information centers. Libraries and other sci-tech information activities for the most part were organized to support specific agency and departmental programs, rather than to carry out any coordinated federal sci-tech information program. For most of its history then, the United States had no overall federal policy in this field.

Some positive actions were taken, however, to consolidate library collections and specialized information activities. For example, at an early date, the Smithsonian Institution began to organize a sci-tech library by purchase or by exchange of materials with other domestic and foreign learned institutions. These actions were in response to a congressional directive that was included in its enabling legislation in 1846. By 1865 the Smithsonian had built one of the largest sci-tech libraries in North America. At the suggestion of the secretary of the Smithsonian Institution, Congress in 1866 authorized transfer of this collection to the Library of Congress; two employees also were transferred to care for and render service on the documents. Congress ordered that this collection be kept separate and that it be available to scientists and engineers in the federal government and to nonfederal scholars.

Preceding the above actions by six years was a congressional ruling that both copies of materials deposited for copyright should go to the Library of Congress. Prior to this time the Smithsonian Institution had received one copyright copy. Also, at that time Congress authorized the Smithsonian Institutions's National Museum (now the Museum of Natural History) and decreed further that all federal collections of fine arts and natural history in Washington D. C. be placed in the National Museum. In 1848 Joseph Henry, the first secretary of the Smithsonian Institution, initiated a national weather service, which began to issue a daily weather map. The Smithsonian enlisted the cooperation of the

Army and Navy in furnishing weather data from their widely scattered establishments and from private individuals and organizations. By 1866, this service proved too heavy a burden on the Smithsonian's private funds, and it was transferred to the Army Signal Corps.

In the 1870s Congress became concerned over the competition among several federal exploration and surveying expeditions in western United States. In 1877 Congress requested the National Academy of Sciences to review the federal projects in this area. The Academy, in 1878, recommended that these expeditions, the Coast Survey, and some of the General Land Office's activities be consolidated into one agency. Neither Congress nor the executive agencies were willing to take such aggressive action at that time, but Congress did abolish three of the four western surveys and gave the responsibilities to the Geological Survey, which it established in 1878. This step led to the development of the Geological Survey's large collections of information on land, water, and other resources of the United States, including an outstanding library and map collection. Another example of federal coordination of sci-tech information services is the Army Surgeon General's Library, which evolved into the medical library for all military services and later into the National Library of Medicine.

The above examples are exceptions rather than examples of general practice. For the most part, federal agencies and departments established sci-tech information services to serve specific programs. Within the Department of Interior, during this period, many of the bureaus developed their own libraries and other information services, including those in the Bureau of Mines, the General Land Office, the Fish and Wild Life Service, and the Geological Survey. In addition the department had libraries in its regional offices. No attempt was made to coordinate these until after World War II. The military departments, the Department of Commerce, the health and social security agencies, and many other independent agencies followed a similar development and operational pattern for sci-tech information services. The Department of Agriculture was an exception; its library was originally organized as a service to most of its bureaus. The department also developed early cooperative activities with the departmental regional centers and land-grant institutions.

We should not assume that the federal government failed to develop any important and valuable information resources during this 150-year period. By 1940 the Library of Congress had grown to be one of the great national libraries of the world. Other federal libraries and information services became important national and international sci-tech information resources; these included the library and information collections of the Patent Office, the Geological Survey, the Bureau of Standards, the Bureau of Mines, the Weather Bureau, and the Department of Agriculture's library and collections on forests, conservation, and agricultural economics. Most of these provided services that included information on countries throughout the world. During this interval the federal government expended tremendous financial and other resources on libraries and other information services. However, each of these information activities was enlarged, reduced, or eliminated depending upon the parent agency's particular programs and not in relation to total governmental or national needs. Consequently, the beginning of World War II found the federal government's sci-tech information inadequate in many areas and completely lacking in some. The following review of selected departmental and agency information activities during this period provides a perspective of federal policies and programs in sci-tech information at that time.

PATENT OFFICE[1]

The Patent Office is an example of a response to a demand for a national service. When the federal government was established, a number of people in the United States wished to protect their inventions. Recognizing this need, Congress passed the Patent Act of 1790, which provided for a governmental committee of experts to examine and pass on applications for patents. This informal system operated until 1836 when the load of patent applications required a more formal system. (In 1837, 436 applications for patents were submitted.) In 1836 a revised patent law authorized an Office of Commissioner of Patents to issue patents. The commissioner of patents was appointed by the president with the consent of the Senate. The commissioner, as well as a Committee of Experts, needed to consult domestic and foreign patents as well as published sci-tech literature in order to establish the novelty of claims. One of the first acts of newly appointed Commissioner of Patents Henry L. Ellsworth was to organize a file of domestic and foreign patents and to assemble a sci-tech library. The Patent Office also began a collection of models, which had to be submitted with patent applications at that time, and some of the geological, botanical, zoological, ethnographical, and other types of specimens collected by federal expeditions were soon placed in the custody of the Patent Office as well, because of its experience in organizing and maintaining such collections. This practice continued until 1857 when these specimens were transferred to the National Museum with an appropriation for maintenance.

The Patent Office was also the repository for specimens of agricultural plants and seeds not found in the United States, which were brought back by expeditions and individual travellers from all over the world. In 1831 Congress authorized a seed distribution program and, soon the Patent Office was distributing thousands of packets of seeds to U. S. citizens (30,000 packets in 1841). Eventually the Patent Office, with congressional approval, established a separate section on agriculture, which employed chemists and biologists to study diseases and insects that were damaging U. S. crops. This section, which by 1839 had accumulated a large collection of materials on agricultural subjects, eventually became the nucleus of the Department of Agriculture Library. The popularity of the agricultural activities influenced Congress to expand support for the section, and in 1862 the section, the seed distribution program, and the library were transferred to the new Department of Agriculture. The chief of the Agriculture Section of the Patent Office, Isaac Newton, became the first commissioner of agriculture. At the suggestion of the Patent Office, Congress included in the 1850 Census authorization for a separate agricultural census that has been continued and refined in later censuses.

In 1860, 4,778 applications for patents were received. The commissioner of patents and his staff were engaged in developing a classification system for patents, as well as in searching for new methods, including mechanization, to handle their rapidly increasing workloads. One of the systems that the Patent Office studied employed punchcards for organizing numerical and other data. This punchcard system[2] was developed by Herman Hollerith at the suggestion of J. S. Billings, who was in charge of organizing the vital statistical data for the 1860 Census. The Hollerith punchcard machine system was first employed in organizing the 1890 Census data. Interestingly, the census staff found the punchcard system useful, but the Patent Office did not actually employ it until after World War II, even though they continued during the pre-World War II period to experiment with machines and other information handling techniques.

In sum, although the Patent Office was founded by Congress as a federal agency to issue patents, it became an archive for many federal collections; developed a large sci-tech library, part of which became the nucleus of the Department of Agriculture Library in 1863; began exploration of techniques for organizing and disseminating information; and initiated a biochemical research program oriented toward aiding American agriculture.

LIBRARY OF CONGRESS[3]

As in the case of patent claims, founders and organizers of the new federal government soon needed a library for reference, legal information, and other purposes. When the government moved to Washington, D. C., it established a library under the supervision of a Joint Committee of Congress on the Library. No one at that time, nor during most of the nineteenth century, envisaged Congress developing a great research library and cultural center. However, a series of isolated policy decisions by Congress led to acquisition and organization of outstanding sci-tech collections. To make these collections useful, Congress provided staff for organizing and servicing them.

During the War of 1812, the Library of Congress was destroyed when the Capitol was burned by the British. A vigorous debate ensued in Congress regarding replacement of the burned library. Many members wished to limit it to a small reference collection with service tailored to the immediate needs of congressional members and committees. Others opted for accepting Thomas Jefferson's library, which he had offered to sell to Congress. Jefferson had assembled one of the most important personal scholarly libraries in the United States. Because of his interest in science, his collection was particularly strong in that field. Congress finally voted to accept Jefferson's offer even though his library was broader in scope than was needed by Congress.

One result of the acquisition of Jefferson's library was the publication in 1815 of a classified catalog of the Library of Congress collections. George Watterson, then librarian of Congress, slightly modified Jefferson's classification system that was based on the Baconian classification of science. Later, in 1864, Ainsworth R. Spofford, when he became librarian of Congress, began reclassifying the collections. The pragmatic classification, which Spofford adopted, organized the materials by topics and form of publication. This system, with refinements, is used today by the Library. By 1896 the Library of Congress collections had been reclassified and were ready to be shelved in the new building according to the Spofford system. Figure 2-1 shows Jefferson's classification system with modifications as used by the Library of Congress. Since most large library collections during the nineteenth century were organized by size and type of material or by date of acquisition, the Library of Congress classification by subject was a marked innovation. Later in this chapter mention is made of the Library's catalog card distribution programs that encouraged the widespread use of the Library of Congress classification system.

In 1840 Congress authorized the Library to initiate direct international exchanges. Receipts from this program, plus the increasing number of federal publications, added considerably to the early science and technology collections of the Library of Congress. Later in the 1840s, when the problem of publishing the results of the Wilke's scientific expedition to the Pacific area faced federal officials, a decision was made to place this publication program under the administration of the Joint Committee on the Library. Thus, the concept that Congress's library could be a center for publication of scholarly treatises and journals, as well as a repository for books and periodicals, was introduced and has never been seriously challenged by Congress since that time. This publishing activity provided the impetus for the Library of Congress to become a repository and information center for cartographic and geographic information and eventually to house one of the world's great collections of maps, charts, and atlases.

Other nineteenth-century congressional decisions continued to enlarge the Library's responsibilities in science and technology. It was designated as the agency to administer the Copyright Acts of 1865 and 1870. As a result, previous material deposited for copyright was transferred to it from the Patent Office in the Department of Interior. From that time to the present, copyright deposits have been a significant source of monographs, serials, photographs, motion pictures, and newspapers that make up the Library's total holdings.

BOOKS may be classed according to the faculties of the mind employed on them: these are—
I. MEMORY.   II. REASON.   III. IMAGINATION.
Which are applied respectively to—
I. HISTORY.   II. PHILOSOPHY.   III. FINE ARTS.

					Chapt.
I. HISTORY	Civil	Civil Proper	Antient	Antient History	1
			Modern	Foreign	2
				British	3
				American	4
		Ecclesiastical		Ecclesiastical	5
	Natural	Physics		Natural Philosophy	6
				Agriculture	7
				Chemistry	8
				Surgery	9
				Medicine	10
		Nat. Hist. Proper	Animals	Anatomy	11
				Zoology	12
			Vegetables	Botany	13
			Minerals	Mineralogy	14
		Occupations of Man		Technical Arts	15
II. PHILOSOPHY	Moral	Ethics		Moral Philosophy / L. of Nature & Nations	16
		Jurisprudence	Religious	Religion	17
			Municipal / Domestic	Equity	18
				Common Law	19
				Law Merchant	20
				Law Maritime	21
				Law Ecclesiastical	22
			Foreign	Foreign Law	23
		Oeconomical		Politics / Commerce	24
	Mathematical	Pure		Arithmetic	25
				Geometry	26
		Physico-Mathematical		Mechanics / Statics / Dynamics / Pneumatics / Phonics / Optics	27
				Astronomy	28
				Geography	29
III. FINE ARTS	Architecture			Architecture	30
	Gardening			Gardening	31
	Painting			Painting	
	Sculpture			Sculpture	
	Music			Music	32
	Poetry			Epic	33
				Romance	34
				Pastorals / Odes / Elegies	35
				Didactic	36
				Tragedy	37
				Comedy	38
				Dialogue / Epistles	39
	Oratory			Logic / Rhetoric / Orations	40
	Criticism			Theory	41
				Bibliography	42
				Languages	43
Authors who have written on various branches				Polygraphical	44

Figure 2-1. Jefferson's classification scheme as printed in the Library of Congress's 1815 catalogue. Courtesy of the Library of Congress

In 1866 the Library of Congress was authorized to accept the scientific library of the Smithsonian Institution, which amounted to about 40,000 volumes. In addition, two employees from the Smithsonian Institution were assigned to

The Beginnings: Federal Information Services, 1790-1942   15

the Library to assist in maintaining and servicing the collection. Provisions in this agreement allowed Smithsonian employees and the public to use the collection. Since the Smithsonian had exchange agreements with scientific institutions throughout the world, the Library of Congress rapidly became internationally eminent in sci-tech publications. In 1869, when the Brussels Treaty was approved by the Senate, the Library of Congress became the recipient of the official publications of many foreign governments through official exchange agreements and treaties executed by the Department of State. The Brussels Treaty was an agreement among seven nations to exchange their official documents, including scientific and other learned treaties.

The next important move by Congress regarding the Library of Congress was to divorce its direct management from the immediate supervision of the Joint Committee of Congress on the Library. This decision was made in two parts. First, Congress authorized and appropriated funds for construction of a separate building to allow removal of the Library from the Capitol Building and thus create space for expansion of the collections and services. The building was completed in 1897. The second part of the decision was to authorize the Librarian of Congress to develop and administer the Library with only general policy guidance by the Joint Committee. Congress, of course, continued to exert basic control through the appropriation process. In 1898 the librarian reorganized the Library of Congress, and among the new units established was the Smithsonian Deposit, to which all sci-tech material in the collections was added. The second unit was the Hall of Maps and Charts, which eventually became the Division of Geography and Maps.

In 1901, Librarian Herbert Putnam received authorization to prepare, print, and sell catalog cards for library materials. These catalog cards gave librarians and scholars ready-made tools for use in their own bibliographic work and as mentioned earlier, also led to wide adoption of the Library of Congress classification code for organizing library materials. Later the same year Congress authorized the Library of Congress, among other government agencies, to make its materials available to scholars and institutions of learning. The Library of Congress continued to expand its services and collections during the next four decades but no further program or policy decisions on its sci-tech collections and services were made until 1947. These actions are discussed in Chapter 3.

In sum, during this period the Library of Congress acquired and organized a preeminent sci-tech collection; developed a widespread network of exchanges of publications; markedly influenced U. S. bibliographic techniques by introducing to many libraries the concepts of subject classification and the dictionary catalog through distribution of its printed catalog cards; and made its materials more readily available through the use of interlibrary loans and the introduction of a photoduplication service. Probably most important to Congress was the initiation of an information and research service to members and committees of Congress, known as the Legislative Reference Service, which is now called the

Congressional Research Service. This action demonstrated to Congress that a library could be more than a place to get library materials: it also could identify and organize information in such a way that it could easily be applied to a given problem.

SMITHSONIAN INSTITUTION[4]

The history of the Smithsonian Institution illustrates a different approach to the development of information resources and services. The administrators of the institution were anxious that it be a center of scholarly research and an aid to scientific publication. Early in its history, however, it became involved in library activities, as well as in building and maintaining a large museum and scientific information collections and has thus had an important role in the development of U. S. science and technology and an even greater impact on scientific information activities. Early projects initiated by the Smithsonian led to the development of the large sci-tech collections of the Library of Congress; establishment of major museums and fine arts collections; stimulation of research and publication of early American science; establishment of the National Weather Service; development of international exchanges of learned publications; initiation of investigations into the fisheries in the United States; and encouragement of early work in aeronautics and astronomy.

In 1826 James Smithson, a British chemist, prepared a will in which he bequeathed his assets, which included about one hundred thousand pounds, to his nephew or the government of the United States if his nephew died without heirs. This bequest was to establish a Smithsonian institution in Washington, D. C., ". . . for the increase and diffusion of knowledge among men." In 1835 Smithson's nephew died without heirs, and the U. S. federal government was notified of the gift. Congress voted in 1836 to accept the funds, but not until 1846 did it establish an organization to use them. This delay was due to debates in Congress about the character of the proposed institution. Some members of Congress wanted to establish a university, others opted for a national museum, some wanted an astronomical observatory, and still others favored a national library.

The provisions of the organic act indicate the compromise that was reached within Congress. It transferred to the Smithsonian all objects of art and natural history in Washington, D. C., that belonged to the federal government; specified an appropriation from the interest that accrued from Smithson's bequest, not to exceed an average of $25,000 a year, be made available for a library; directed that one copy of all copyrighted publications be deposited in the Smithsonian library and one copy in the Library of Congress; and authorized the managers of the institution to spend the income from the Smithsonian fund as they deemed best to promote the purposes expressed in the will.

Joseph Henry, an eminent U. S. physicist and the first secretary of the Smithsonian Institution, immediately began construction of the building on the site selected by Congress and in 1846 outlined his first program. Under the provision of the will for the "increase of knowledge," it included support for research projects and publication of the results in a series known as "Smithsonian Contribution to Knowledge." (This series was continued for sixty-eight years and did much to encourage the development of American science.) Under the provision for the "diffusion of knowledge," the program called for periodical reporting on the progress of science in the various branches and support for the preparation and publication of separate treatises on scientific subjects of general interest. Although James Smithson's will gave no indication of the fields of learning the institution should emphasize, Joseph Henry's background and interests in science were reflected in his program for the Smithsonian Institution as well as in the early projects initiated by it.

Henry was apparently unhappy about the provision in the act for the development of the Smithsonian library because he thought this responsibility should belong to other institutions, such as the Library of Congress, in 1847 he did appoint an assistant secretary, Charles Coffin Jewett, who was to be responsible for the library. Jewett was anxious to build a strong library for research and almost immediately disagreed with the secretary on the scope and size of the library operations and collections. Assistant Secretary Jewett began to develop exchange agreements with learned organization throughout the world and zealously acquired valuable scientific publications from within the United States. After eight years the differences between Jewett and Henry became so great that Jewett left the Smithsonian.

In 1860 Henry was instrumental in having a section of the Copyright Law changed so that both copies of new publications went to the Library of Congress, and in 1866, partially because of a disastrous fire the previous year, the Smithsonian transferred the bulk of its library collections to the Library of Congress. Congress concurred in the move but decreed that these materials should be readily available to the staff of the Smithsonian and could also be used by the learned public. The Smithsonian has continued to maintain a working library but when the bulk of its collections went to the Library of Congress, it also transferred to the Library of Congress the means to build broad scientific and technical collections for the federal government as a whole. One direct outcome of Jewett's efforts to build a Smithsonian library was the growth of extensive exchange agreements for learned publications. The Smithsonian in 1852 established the International Exchange Service that was principally a receiving and shipping center. The service was made available to other U. S. learned institutions. Congress gave support to this activity by granting it free mailing privileges when the service was started, and in 1887 Congress directed that all printed congressional and executive agencies' documents should be available for exchange. In 1881, when the International Exchange Service became a financial burden to the

Smithsonian, Congress appropriated funds to support the services and has continued to do so. However, these appropriations have never been sufficient for the International Exchange Service to give prompt service; nevertheless it has and continues to be a useful aid to government agencies' and learned institutions' exchange programs.

As indicated earlier, many appointed and elected federal officials wanted the Smithsonian Institution to serve as a national museum. This desire was reflected in the provision of the enabling act that transferred all federal collections in natural history and art to the Smithsonian. Prior to Smithsonian's bequest, Congress in 1818, had chartered the Columbian Institute for Promotion of Arts and Science. One purpose of this institute was to collect and study various minerals and natural curiosities of the United States. The institute's charter was not renewed in 1838, but in 1840 Congress chartered The National Institution for the Promotion of Science. Several federal collections including the collections of the Columbia Institute as well as Smithson's personal collection of minerals, which had been placed in its custody, were transferred to the Smithsonian Institution when it was established. Even though Secretary Henry clearly did not favor this activity, in 1850 he appointed Spencer Fullerton Baird, a young zoologist from Carlisle, Pennsylvania, as assistant secretary with responsibility for such material. Baird also brought to the Smithsonian his own large and varied collections of birds, mammals, plants, fossils, and other objects of natural history and, as assistant secretary of the Smithsonian, rapidly developed large and varied collections of art and natural history objects. Again, Secretary Henry objected to Smithsonian resources and space being utilized by an activity he considered outside the Smithsonian's prime mission. However, he acquiesced to congressional interest, and in 1857 federal collections in the Patent Office were moved to the Smithsonian and Congress began annual appropriations for the support of the Smithsonian's National Museum. The museum activities fostered by Baird, first as assistant secretary and later as secretary of the Smithsonian, laid the base for a number of the Smithsonian's important programs, such as the National Fine Arts Collection, the Museum of Natural History, and the Museum of History and Technology.

In 1848 Joseph Henry initiated a service that has become one of the country's most important information gathering and dissemination programs. He proposed a cooperative venture in which daily weather observations would be taken in various parts of the nation, sent to the Smithsonian, and used in the preparation of a daily weather map. With cooperation from individual observers, army posts, and naval establishments, the Smithsonian soon developed a National Weather Service that proved useful to agriculture, business, industry, and the military. The invention of the telegraph aided greatly in the effectiveness of this service and also permitted the rapid exchange of information about storms and other unusual weather and meteorological phenomena among widely separated observers. The data collection and organization of the National Weather Service so taxed the resources of the Smithsonian, however, that in 1869 the service was transferred to the Army Signal Corps. One outgrowth of this Smithsonian project was initiation of the compilation of meteorological tables by Professor Guyot of Princeton University. These and the comprehensive weather records organized by the Smithsonian started a data base that has been immensely important to meteorological and climatological research in the United States.

In 1870 and 1871 the Smithsonian Institution began two federal programs that have contributed significantly to the advance of American science by developing large and valuable information resources. The Bureau of American Ethnology, established by congressional action in 1870, was the outgrowth of interest in American Indian culture aroused by expeditions in the early part of the nineteenth century. Many explorers brought back fossils and other specimens related to the Indian tribes that they encountered in their travels. John Wesley Powell, who became interested in Indian life and culture on his first trip to the West, was appointed as the first head of this bureau. Both Secretary Henry and Assistant Secretary Baird encouraged and supported work in ethnology. Their encouragement, along with Powell's vigor and effective work, made the bureau an outstanding research center in anthropology and ethnology. When Powell was appointed director of the U. S. Geological Survey, he continued to serve as the head of the Bureau of American Ethnology and returned full time to the bureau when he retired from the Geological Survey. An effective cooperative program developed between the bureau and the Smithsonian's National Museum, whereby the bureau concentrated on acquisitions, research, and publication, and the museum organized and maintained the collections. Among the noteworthy publications resulting from the Smithsonian program are *The Handbook of American Indian Languages* (1911), *The Handbook of Indians of California* (1925), and the *Handbook of South American Indians* (1950).

In 1871 the personal interest of Assistant Secretary Baird led to the establishment by Congress of the U. S. Fish Commission, with Baird as its chairman. Baird and others were concerned about the depletion of fish in U. S. coastal waters, lakes, and streams, and under Baird's direction much was done to save the salmon, shad, white fish, and carp industries of the United States. The results of oceanic expeditions were documented, and specimens classified and preserved in the Smithsonian. In 1941 the commission's program was incorporated into the Fish and Wildlife Service of the Department of Interior, but the specimens remained a part of the Smithsonian collections.

In 1865, the Smithsonian published a general catalog of scientific papers that became an important bibliographic tool of librarians and scientists throughout the world. The Institution found, however, that maintaining this project placed too heavy a burden on its staff and financial resources. Consequently, at the Smithsonian's suggestion, two international meetings were held to develop a cooperative program to continue this bibliography. In 1901 the participating countries agreed to prepare annual card catalogs of their respective publications and to forward these to the Royal Society in London for consolidation and publication. Two hundred and fifty volumes of the *International Catalogue of Scientific Literature* were published between 1901 and 1915. The lack of resources and the impact of World War I ended this cooperative international project. The need for scientific bibliographies remained, of course, and such English-language specialized bibliographic tools as *Chemical Abstracts, Mathematical Reviews,* and *Science Abstracts* were established. Similar bibliographic services are published in other languages, but no attempt has been made to prepare a comprehensive international bibliography since the early 1900s.

Other areas in which the Smithsonian made significant contributions to scientific research and supporting information services during the early part of the twentieth century include aeronautics, geophysics, and astronomy. For example, the Institution was a party to formation of the National Committee for Aeronautics in 1915. Immediately after World War II, the Smithsonian embarked on an information program that is discussed in Chapter 3.

COAST AND GEODETIC SURVEY[5]

The missions of the antecedent agencies of the Coast and Geodetic Survey were to chart the territorial waters of the United States and to develop aids to navigation. These programs led to work in geodesy and geophysics. The expanded program was recognized in name when the Coast and Geodetic Survey was established in 1878. The activities of this agency have resulted in large information files and services.

The beginning of the Coast and Geodetic Survey dates back to the beginning of the nineteenth century. Even before the establishment of the U. S. federal government, commercial navigation, fishing, and naval requirements for better information on coastal and esturial waters, coastline configurations, and navigational aids made hydrographic surveys necessary. Charts were prepared by private individuals and by governments involved in the exploration and administration of the eastern coast of North America. In 1802, after much debate and some spasmodic surveying and charting by federal agencies, Congress began appropriating funds for a coastal survey. This effort was viewed as a short-term, definitive project. By 1807, demands from private marine and naval organizations encouraged Congress and the executive branch to advertise for a plan to survey accurately the east coast of the United States. Ferdinand Rudolph Hassler, who was employed to develop standard weights and measures for the U. S. Customs Office, submitted the successful plan and was appointed the director of the project called the "Coast Survey." Hassler's main interest was to establish a high-grade scientific organization that could produce work comparable to the best in Europe. The methods of measurement and triangulation introduced in the surveys conducted during his term of office were a mix of the ideas of the best civilian and military scientists and engineers. However, his decision to delay publication of charts and other navigational aids based on these surveys since he was not satisfied with their quality was an action that disappointed both Congress and his superiors in the Treasury Department.

In 1843, Alexander Dallas Bache assumed directorship of the Coast Survey. Aware of the criticism of the delay of publication and the inefficient management during the earlier period, Bache organized eight field teams to conduct surveys and quickly began to publish the results of these and the previous surveys. In addition to charting coastlines, Bache continued the earlier work on tables of magnetic variations, establishment of accurately determined base lines, and studies of tides. He initiated studies of the Gulf Stream and sea bottom sediments and was successful in persuading Congress that the Coast Survey programs should be long term in order to continue needed coast and geodetic work on the continental United States and its expanding territories. Some actions by Bache as director of the Coast Survey, however, resulted in its being a target for Navy Department criticism. For example, aware of the necessity for continuous astronomical surveys for sailing and other navigational and charting purposes, he turned to Benjamin Pierce at Harvard College for this work even though the Naval Observatory under Matthew F. Maury was available.

During both Hassler's and Bache's regimes the Coast Survey was transferred temporarily to the Navy Department. Following the Civil War, the Coast Survey was expanded under the Treasury Department and in 1878 became the Coast and Geodetic Survey. It continued programs of geodetic, tidal, and coastal surveying. The acquisition and organizations of these data files and publication of results laid the base for the large information resources that the Coast and Geodetic Survey has at the present time. Later activities of this agency and its incorporation into the National Oceanographic and Atmospheric Agency is the subject of further discussion in Chapters 3 and 4.

GEOLOGICAL SURVEY[6]

The U. S. Geological Survey was established in 1878 as the result of congressional concern about the competition among four on-going major surveys in the West, and this agency became one of the most significant federal establishments for gathering, organizing, and disseminating information on the nation's geography, geology, water, land, and mineral resources. The four surveys that were the forerunners of the Geological Survey were the geological survey of the 40th parallel under the sponsorship of the Army but headed by a civilian, Clarence King; the Army's military and geographical survey west of the 100th parallel under George M. Wheeler; the geographical and geological survey of the territories by the General Land Office, led by Ferdinand Hayden; and the survey of the Colorado and Rocky Mountain areas directed by John Wesley Powell under the sponsorship of the Smithsonian Institution.

In 1877 Congress was concerned about the overlapping activities among these surveys, and asked the National Academy of Sciences to review the situation and make recommendations. In 1878 the Academy recommended that one agency, the Coast and Interior Survey, be established in the Department of Interior. In addition, it urged major changes in the system of land offices and land surveys. The recommendations were never fully accepted, but they were used as a basis for later congressional action. In 1879 the Geological Survey was established in the Department of the Interior by a rider on an appropriation bill. Of the four on-going surveys listed above, all but King's were abolished and their responsibilities transferred to the Geological Survey. King was appointed director and began immediately to recruit a staff and organize a program. King intended to remain in office only as long as necessary to develop and set in operation plans for a scientific agency. Although he resigned in 1881, during his two years as director, King began many activities that led to the Geological Survey's becoming a great information source. His annual report for 1880 listed twelve titles of forthcoming publications; he initiated a cooperative program with the Census Bureau to compile and issue yearly statistics on minerals; he began a series of topographic and geologic maps that would cover the United States; and he started a study of geological classification and nomenclature.

John Wesley Powell, the second director, built on the preliminary work of King. He initiated a series of topographic maps of the United States at scales of 1:250,000, 1:125,000, and 1:62,500. The series included 2,600 sheets covering the United States; also in the plan were several hundred large-scale special maps of areas of high interest such as mining, irrigation, and forestry, where more detail was necessary. These map series have provided basic data used in education, research, commerce, and industry. Powell also organized work on the geologic map of the United States and begun to issue annual reports (with special studies attached as appendixes), bulletins, and monographs. In 1894 a series of geologic folios with maps and text was started. In order to have necessary published resources available for its own research and publication program, the Geological Survey organized a library, which in 1882 included 25,000 volumes and maps.

Director Powell, as well as his predecessor, recognized the importance of bibliography and classification. In 1882 he encouraged the Washington Philosophical Society to consider the problems of geological classification. One of the results of these deliberations was a continuing bibliography of American and foreign geology, which began publication in 1883 and has continued in different forms and under different sponsorship to the present time. Powell continued King's work on geological nomenclature and classification at both the national and international levels. Correlations of geological terms and geological dating were an important element of this work.

In sum, by 1942 the U. S. Geological Survey had produced topographic and geologic maps covering a major portion of the United States; issued hundreds of detailed maps of important mineral and other natural resource regions; started long-term hydrological surveys of the streams of the United States; developed geological classification; and organized a major earth science library and map collection. In addition, it had organized valuable data files on U. S. earth resources.

EARLY MILITARY PROJECTS[7]

The new U. S. government turned to the military services not only for defense but also to carry out many nonmilitary projects. This practice led to the establishment of information gathering and disseminating units that were the forerunners of some of today's major scientific technical and medical information resources and services. Examples of these include the Hydrographic Office, the Naval Observatory, the National Library of Medicine, and the Defense Topographic Center.

The Army, particularly the Corps of Engineers, was involved in most of the early governmental exploratory and mapping expeditions in the continental United States. Surveys of the Great Lakes and major streams and esturaries were also the province of the Army's Corps of Engineers. The names of Lewis, Clark, Long, Pike, Wheeler, King, Powell, and Fremont are associated with major expeditions that gathered important data and other information. These men either were military officers or, like Powell, leaned heavily upon military support. Most of the reports, specimen collections, and maps that resulted from their expeditions and surveys were transferred to civilian agencies, such as the Smithsonian Institution, the Library of Congress, the Patent Office, and the Geological Survey for safe keeping and exploitation. The Army Medical Department began the gathering of weather records when it was established in 1818 for possible medical purposes,* and these records were furnished to the Smithsonian when it developed the National Weather Service.

Other than the weather data, few medical records or research data were maintained by the Army Medical Department until the latter part of the nineteenth century. In 1836 an early Army surgeon general, Joseph Lovell, did organize a library in his office, which was augmented in succeeding years such that a catalog published in 1864 listed 1,365 volumes. In 1865 John Shaw Billings, a young medical officer who was placed in charge of the surgeon general's library, developed it into an outstanding institution. He was the originator of the *Index Medicus*, a bibliography of medical publications, and other services, and he contributed significantly to the organization and program of the Army Medical Museum. Associated with the library and museum were George M. Sternber, whose studies in bacteriology and especially on yellow fever are noteworthy, J. J. Woodward, who contributed significantly to the use of microphotography in the study of bacteriology; Walter Reed, whose work on typhoid

---

*Available information does not make clear why the Army began this program.

and yellow fever is well known; and W. C. Gorgas, who also worked on yellow fever and malaria in the tropics and subtropics. This library was the forerunner of the National Library of Medicine, which is discussed in the next three chapters.

The Navy Department also made important contributions to the early development of information resources and services that later became national information centers. Because of the Navy's need for navigational data and instruments, the secretary established the Depot of Charts and Instruments in 1830. Before that time commanders of naval vessels had to procure their own charts and navigational instruments. In 1833 the first naval observatory was established with Charles Wilkes as director. By 1845 the observatory had a continuing program for making observations on the sun, planets, and bright stars. In 1848 Congress approved a program of meteorological observations; and in 1856 the present U. S. Naval Observatory was constructed and Congress appropriated funds for equipping and staffing it.

From the beginning, the Navy engaged in hydrographic surveys and in preparation of hydrographic and sailing charts and sailing directions. In 1835 it acquired a lithopress, and the first four engraved charts were printed in the United States in 1837. In 1867 the Navy began to print *Notices to Mariners* and notes on hydrographic features such as shoals, reefs, small islands, and so forth as part of the output of the new Navy Hydrographic Office that had been established by Congress in June 1866. (Incidentally, this agency was first housed in the famous Octagon House.) In 1883, the Hydrographic Office began to issue pilot charts for the North Atlantic and in 1894, for the North Pacific. These charts were based in part on Maury's wind and ocean charts that were issued between 1844 and 1861.

Just as the Army explored the continental United States, the Navy very early was given responsibilities for exploration and scientific investigation of the seas. Specific ventures include Wilkes's expedition to the Pacific in the 1840s, Gilless's naval astronomical expedition to Chile in 1849, and various expeditions to the Amazon and the La Plata in the 1850s. As in the case of the Army, the resulting data, maps, charts, and specimen collections were often transferred to civilian agencies.

In sum, by the beginning of the twentieth century the military departments had established agencies whose information activities were the basis for national information services and resources in topography, hydrography, meteorology, climatology, geology, geography, geodesy, marine biology, and medicine. Further military sci-tech information activities are reviewed in Chapters 3, 4, and 5.

DEPARTMENT OF AGRICULTURE[8]

Development of the Department of Agriculture Library and specialized information resources differs from that for other departments in its initial emphasis on a centralized library system with coordinated information services and data files. Since early in the nineteenth century, American agriculture has been the recipient of innovative federal information and assistance policies, including information about and distribution of new and improved seeds and plants. Cooperative research and demonstration programs with states and counties, the building of a major national agricultural library, and the preparation of technical bulletins on myriads of topics has also assisted U. S. agriculture.

As noted earlier, by 1831 the agriculture component of the Patent Office was of sufficient size to be a separate unit, whose activities continued to expand during the next three decades. By 1860 there was a movement for a separate federal agricultural agency, and in 1862 Congress authorized a Department of Agriculture, although it was not given cabinet rank until 1898. The enabling legislation included authority, ". . . to acquire and diffuse . . . useful information on subjects connected with agriculture . . . procure and propagate seeds and plants." During the same year the Morrill, or Land-Grant, Act was passed ". . . donating public lands to several states and territories which may provide colleges for the benefit of agriculture and mechanic arts." These land-grant colleges were the forerunners of many eminent state universities. In 1905 the Forest Service was established as a separate unit under Gifford Pinchot. Before this date, forestry was a subunit in the departmental structure. The service immediately began to expand the program for inventorying, conducting research on, and improving management of forest areas of the United States. Thus began the organization of large information resources on forests and forest products, which led to the establishment in 1910 of the Wood Products Laboratory in Madison, Wisconsin. This organization was the forerunner of several specialized regional research laboratories. In 1906 the Department of Agriculture was authorized to monitor the purity and quality of foods and drugs, and the Bureau of Chemistry was given this responsibility.

This brief summary of the Department of Agriculture's major programs indicates the wide scope and varied information activities that were developed to give support of its basic mission. All of these activities had a direct impact upon the character and growth of the departmental library. In 1864 the department's appropriation included $4,000 for the salary of a librarian and for purchase of laboratory equipment. The library was formally organized as soon as the new agriculture building was completed in 1868. By 1872 its collections numbered 8,000 volumes and exchanges had been arranged with agriculture organizations in many foreign countries and all the U. S. states. In addition, agricultural experts (called "plant explorers") who were formally or informally connected with the department were commissioned to search for plants with agricultural potential in foreign areas having climatic and soil conditions similar to those in various regions of the United States. These experts collected many publications on agriculture and allied subjects to be added to the library, although those in Oriental and other languages that could not be read by the Department of Agriculture staff were transferred to the Library of Congress.

By 1892 the agricultural department library had a staff of six persons and its collections had grown to 59,000 books and pamphlets. In 1880 the library prepared and printed a catalog of the departmental publications that enhanced the reputation of the library and enabled it to expand its exchange arrangements. As the department expanded its program to include forestry, animal husbandry, soil conservation, pure food and drug administration, and agricultural and home economics, the library services and collections were correspondingly enlarged to support these programs. By 1912 the library had a staff of thirty-nine, the collections numbered 122,000 books and pamphlets, and 2,000 periodicals were being received. It had become one of the great agricultural libraries of the world. Prior to this date, in 1902 the Department of Agriculture Library began to cooperate with the Library of Congress in the latter's printed catalog card distribution program. This joint effort enabled the land-grant colleges, agricultural experiment stations, and regional laboratories to receive information on current publications in agriculture and allied subjects. In addition the department distributed its own publications and loaned other unusual publications to these institutions.

In sum, by 1942 the Department of Agriculture was the focal point for a large, effective information system that has been a significant force in the rapid advance of American agriculture. Important components of this system included the library, the agriculture experiment stations with their publications, demonstration plots and farms, the regional research laboratories, and the large and varied departmental publication programs that produced low-cost books and pamphlets on subjects in a style that appealed to the American public. The Department of Agriculture's information program during this period was one of the more effective ones in the federal government. Most of the department's bureaus relied on its central library. Congress had strengthened the move toward centralization by decreeing that funds for library materials had to be identified in budget requests. The library was the first in the federal government to distribute printed catalog cards of its acquisitions to regional and state libraries. The department was using photocopies of library materials in lieu of loans as early as 1911. The departmental library cooperated in 1936 with Science Service and the American Documentation Institute in developing bibliographic and microfilm services. Effective liaison had been established with state agricultural experiment stations so these could draw upon the information resources at the federal level. The departmental information program was user oriented because Congress indicated from time to time through such acts as the Hughes and Hatch legislation that it intended that the department make strong efforts to apply its research program and information resources to the benefit of American agriculture.

CONCLUSION

The foregoing review of departmental and agency information activities contains only illustrative examples of federal programs and policies that laid the basis for the broad and varied complex of sci-tech libraries and other information resources and services that were extant at the beginning of the World War II period. The following three chapters will examine in more detail the federal government's programs and policies in sci-tech information from 1942 to 1972.

NOTES AND REFERENCES

1. See A. Hunter Dupree, *Science in the Federal Government: A History of Policies and Activities to 1940* (Cambridge, Mass.: Belnap Press of Harvard University, 1957), 460 pp. For patents, see pp. 7-11, 43, 46-47, 111, 113, and 338.
2. See National Library of Medicine, *Historical Chronology and Selected Bibliography Related to the National Library of Medicine* (Washington, D. C.: National Library of Medicine, History of Medicine Division), p. 4, footnote 3.
3. David C. Mearns, *The Story up to Now: The Library of Congress, 1800-1946* (Washington, D. C.: Government Printing Office, 1947), 226 pp., reprinted from the Annual Report of the Library of Congress for the fiscal year ending June 30, 1946, contains a detailed account of the development of the Library during this period.
4. See Paul H. Oehser, *Sons of Science: Story of the Smithsonian Institution and Its Leaders* (New York: H. Schuman, 1949), 220 pp.; and Dupree, *Science in the Federal Government*, pp. 83-90, 130, 131, 221, 235, 237, 251, 283-87, 289, 291, 330, 335, and 377-70, for discussion of the development of the Smithsonian Institution.
5. Dupree, *Science in the Federal Government*, pp. 29, 33, 43, 52, 53, 64, 65, 86, 100-05, 132, 202, and 203, reviews Coast and Geodetic Survey. Also see G. A. Weber, *The Coast and Geodetic Survey: Its History, Activities and Organization*, Monograph No. 16 (Baltimore: Institute for Government Research Service, 1923), 82 pp., which stresses legal and organizational changes.
6. Dupree, *Science in the Federal Government*, pp. 177, 195, 208-12, 217, 226-29, 233-35, 248-50, 275, 280, 281, 299, 304, 326, 339, and 352, discusses the Geological Survey's development. Also see John C. Rabbitt and Mary C. Rabbitt, "The U. S. Geological Survey: 75 Years of Service to the Nation, 1879-1954," *Science* 119 (May 28, 1954): 741-58; and *The U. S.*

28   Two Centuries of Federal Information

Geological Survey: Its History, Activities and Organization, Monograph No. 1 (New York: Institute for Government Research Service, 1919), 96 pp.
7. See W. Stull Holt, *Office of Chief of Engineers of the Army: Its Non-Military History, Activities and Organization*, Monograph No. 27 (Baltimore: Institute for Government Research Service, 1923), 166 pp.; and Dupree A. Hunter, *Science in the Federal Government*, pp. 33, 43, 127, 128, 184-94 for the military in general; pp. 95, 122-26, 133, 134, 218, 53, 56, 57 for the Navy; pp. 26, 28, 53, 63, 92-95, 187, 195, 301, 302, 314, 315, 367, 368 for the Army; pp. 193, 194, 196, 250, 251, 289, 303 for the Corps of Engineers; pp. 128, 129, 256, 263-67 for the Medical Corps; and pp. 192, 193 for the Army Signal Corps. See also Charles Wilkes, *Narrative of the United States Exploring Expeditions During Years 1838, 1839, 1840, 1841, 1842* (Philadelphia: Institute for Government Research Service, 1845), 5 vols., plus Atlas.
8. A. C. True, *A History of Agriculture Experimentation and Research in the United States: 1607-1926*, U. S. Department of Agriculture Misc. Pub. 251 (Washington, D. C.: Government Printing Office, 1937), 321 pp.; Dupree, *Science in the Federal Government*, pp. 111, 113, 149-61, 169-72, 192, 193-244, 254, 348, and 349, for discussion on the research and organization aspects of the Department of Agriculture; and *National Agricultural Library: A Chronology of Its Leadership and Attainments, 1839-1973* (Beltsville, Md.: Associates of National Agricultural Library, Inc., 1974), pp. 1-3.

Chapter 3

Information Centers, 1942-1956

Between 1942 and 1956 the U. S. sci-tech sector experienced marked changes. R&D was greatly expanded and became more diversified, and accompanying this growth was a rapid geographical dispersion of laboratories. The R&D expansion during this period resulted in significant changes in sci-tech information activities. Distribution of printed results of R&D changed from almost complete reliance on traditional journals and monographs to widespread use of scientific and technical reports;[1] information services exhibited the same characteristics of growth, diversification, and dispersion as did the R&D laboratories. Many new for-profit information companies appeared; sci-tech journals increased in size and variety; new technology encouraged the collection and use of sci-tech data; and increased speed in transportation and communications resulted in scientists and engineers demanding more rapid dissemination of the results of R&D. World War II and the following period of political tension, known as the "Cold War," were important stimuli for the expansion, diversification, and dispersion of both scientific R&D activities and the supporting library and other special information activities.

The federal government, in response to the above conditions, established several new types of sci-tech information centers. The major role of some of these was acquisition, announcement, reproduction, and distribution of sci-tech reports; other centers organized current inventories of on-going research and development projects, primarily for the use of R&D administrators; and still other centers served highly specialized segments of the R&D community, such as scientists engaged in cancer research, or those exploring the chemical effects on biological organisms, or investigating the characteristics of metals and other materials. During this period many new private organizations that had important roles in the sci-tech information field were also formed. Among these were Documentation Incorporated, Herner and Company, Eugene Garfield and Associates (which later became the Institute for Scientific Information), Mitre Corporation, the Systems Development Corporation, and organizations associated with universities, such as the Stanford Research Institute and the Johns Hopkins University Applied Physics Laboratory.

29

Among the established federal R&D agencies that were expanded during the 1942-1956 period and given new responsibilities were the National Advisory Committee for Aeronautics, the Weather Bureau, the Coast and Geodetic Survey, the Bureau of Standards, the U. S. Geological Survey, the Bureau of Mines, the Petroleum Production Board, and the Public Health Service. The impact on military R&D agencies was even more marked, and many new organizations came into existence, including the Office of Naval Research, the Office of Scientific Research and Development, the National Science Foundation, and the Manhattan District Project, which later became the Atomic Energy Commission. Most of these agencies relied on the universities and private research organizations to conduct R&D under contract. Therefore, these institutions experience similar growth.

At first, the R&D administrators depended for their information needs upon established libraries and specialized information services. The increased demand, however, overwhelmed the existing services. It also brought to light gaps in library collections and inadequacies for meeting the new and varied information demands of the scientists and engineers and created other problems as well. For example, the expanded use of the technical report presented many problems to librarians, publishers of sci-tech journals, and both the Patent and Copyright Offices. Librarians found that this material presented both custodial and cataloging difficulties. Some journal editors and publishers took the position that technical reports, even though restricted in distribution, constituted prior publication and, therefore, they rejected the same material when it was submitted to their scientific journals. Other editors accepted such manuscripts. When an item that might later be considered for a patent was described in a technical report submitted to fulfill a contract requirement, a question of prior disclosure was raised. Technical reports also presented copyright problems. Some federal agencies took the position that material issued in fulfillment of a government grant or contract should be considered government publication and not eligible for copyright; others did not.

The funding for libraries and information centers frequently did not come from the R&D agency budgets. During World War II, for example, federal libraries were being overwhelmed with demands; yet federal administrators and Congress were reluctant to make increased allocations of funds and manpower to these information agencies. Instead, federal R&D administrators developed or funded new information units closely coordinated with the sci-tech activities under their administration or program control. The following sections review the development of a few of these new information agencies.

DEFENSE DOCUMENTATION CENTER[2]

The antecedents of the Defense Documentation Center date back to 1926 when the Air Corps began to issue a *Technical News Service* at McCook Field, Dayton, Ohio. In January 1929 the *Technical News Service* was transferred to Wright Field where it was continued until 1957 when it was moved to Washington, D. C. In 1945, after the end of the European phase of World War II, Air Force technical intelligence teams consisting of 1,500 technical Air Force personnel combed Germany for sci-tech documents. These were screened at a center in Hanau, Germany. In July of that year the Air Documents Research Center was established in London by Air Force Intelligence. This center, headed by H. M. McCoy of the USAAF, established a cooperative program with the U. S. Naval Intelligence and the British War Ministry. By December the Air Documents Division of the Intelligence Department of the Headquarters, Air Technical Service of the Air Force in Dayton, Ohio, had received over 800,000 documents from the European screening centers. Albert A. Arnhym of the Air Force, who later was to play an important role in military sci-tech information activities, was in command of the Dayton Air Documents Division. In the interests of better coordination, the Navy Bureau of Aeronautics Liaison Office was transferred to Wright Field. After the United States entered Japan in 1945, a similar sci-tech document acquisition program was mounted there; these documents also were sent to the Air Documents Division at Wright Field for cataloging, indexing, translation, and dissemination. In addition to handling foreign documents, the Air Documents Division also distributed U. S.-produced sci-tech reports to Air Force research centers and contractors. This latter activity later led to jurisdictional and technical problems between the Air Documents Division and other civilian and military technical information centers.

Within the United States several major federal projects that were established during the war generated sci-tech reports and data. In 1941 the Office of Scientific Research and Development (OSRD), headed by Vannever Bush, was organized to further the U. S. scientific R&D program in support of the war effort. OSRD's efforts produced some 33,000 documents, which were placed in the custody of the Office of Naval Research (ONR) when it was formed in 1946. Shortly thereafter, in 1947, ONR turned to the Library of Congress for assistance in cataloging, indexing, abstracting, announcing and distributing these and other Navy sci-tech reports. The Library of Congress organized the Science and Technology Project, headed by Mortimer Taube, to meet the requirements of a contract with ONR.

In 1948 the Dayton Air Documents Division, recognizing the need to improve its indexing of documents, contracted with the Institute of Aeronautical Sciences to prepare a standard aeronautical index. The institute established a project in the Los Angeles area to prepare this index under the direction of Leslie E. Neville. Also with Air Force funds, the institute organized a library and technical report information center to serve the aeronautical industry on the West Coast. This action added another center that cataloged, indexed, and distributed sci-tech documents.

Thus the organization of the Dayton Air Documents Division and the Library of Congress Science and Technology Project led to competitive activities. Each of these two units had its own system for cataloging, indexing, and announcing U. S. technical reports. Users in the military and contractor laboratories complained about the duplicative and variant identification of the same technical report. Also, technical report-producing organizations objected to having to furnish their reports to two different federal agencies that served overlapping publics. (Later the Department of Defense took corrective action that ultimately resulted in the formation of the Armed Services Technical Information Agency in 1951.)

In September 1947 the Air Force became a separate military department. Soon thereafter, the Air Force took over the Navy Liaison Office at Wright Field and agreed to serve the Navy Bureau of Aeronautics through the Air Documents Division. That year, the division completed cataloging and indexing 56,000 German documents and 3,000 Japanese documents and also began to collect and service sci-tech reports of interest to the Air Force and the Navy Bureau of Aeronautics. In 1948 the Central Air Documents Office (CADO) replaced the Air Documents Division; it was supported by both the Air Force and the Navy. Arnhym, who was named director of the new office, expanded CADO's program to include government agencies and contractors and carried on an aggressive program of publicizing CADO's collections and services. The next year the Department of the Army joined in the support of CADO. In 1950 CADO enlarged its program still further by establishing a West Coast office in Los Angeles, although the Air Force was already supporting the Institute of Aeronautical Sciences' Library and report center. During this same period the Office of Naval Research and the Department of the Army supported the Library of Congress in expanding its contractual projects on reports and other technical literature. The ONR project in Library of Congress was renamed the Navy Research Section (NRS) in 1949. It had began to issue an abstract bulletin on reports it received, and it instituted reference, loan, and special bibliographic services. NRS was authorized to employ liaison officers who visited users to brief them on how to make the best use of the section's services. CADO had instituted liaison officers' visits prior to this action.

It soon became apparent to the Department of Defense that coordination was essential among these information services being supported by the several military departments. In May 1951, Secretary of Defense George C. Marshall signed an order establishing the Armed Services Technical Information Agency (ASTIA) with Leslie Neville as the first director. This order gave the Joint Defense Research and Development Board responsibility for policy guidance of ASTIA and placed management control in the Research and Development Command of the Air Force. In January 1952, CADO was taken over by ASTIA and renamed the ASTIA Documents Service Center with the major responsibility for furnishing copies of foreign and domestic reports to authorized users. In

1953, the funding and administrative responsibility for the ONR project in the Library of Congress was transferred to ASTIA, and the project was renamed the ASTIA Reference Center. This center was made responsible for the cataloging, abstracting, indexing, and reference services. The Library of Congress continued the preparation of special bibliographies under contract with various units of military and civilian agencies. In 1954 the Air Research and Development Command assumed entire responsibility for the funding of ASTIA, and the above arrangements continued until 1958 when the ASTIA Documents Service Center was moved to Arlington Hall Station in Virginia. The librarian of Congress had suggested to the Air Research and Development Command in 1957 that it also assume operational control of the ASTIA Reference Center and remove it from the Library. This transfer was accomplished at the time the ASTIA Service Center was moved to Arlington Hall and thus the two sections of ASTIA were placed under one operational director. The Library of Congress retained the special bibliographic projects that were funded separately as part of the Technical Information Division of the Library.

From the time ASTIA was established in 1951, it was under pressure from its policy committees (variously named and constituted) to improve and expand its services to the Department of Defense R&D centers and contactors, as well as to other federal agencies. A number of actions were taken to enable ASTIA to be a more responsive technical information center. In 1952 ASTIA was authorized to extend its services to civilian agencies with defense-related activities —such as the Atomic Energy Commission and the National Advisory Committee for Aeronautics—and to establish a second regional office in New York City. In 1956 ASTIA was authorized to provide technical reports with no security or other release restrictions to the NATO nations; in 1958 this authority was extended to SEATO nations. Numerous technological innovations were introduced into the operations during this period, but since these were also being introduced into other information centers, they are discussed separately. Just as an example, however, ASTIA introduced the use of microfilm and microfiche into its storage and distribution operation at the time many federal and private organizations were doing the same. The story of ASTIA's being integrated into Defense Department sci-tech programs and being renamed the Defense Documentation Center, as well as the executive branch's effort toward better coordination among the sci-tech information centers, is related in the next chapter.

NATIONAL TECHNICAL INFORMATION SERVICE[3]

The civilian government agencies also were faced with the problem of the release and distribution of both foreign and domestic sci-tech publications that resulted from wartime and immediate postwar activities. In June 1945, the Publications Board was established by Executive Order 9569. This interdepartmental committee, composed of the attorney general and the secretaries of Interior, Agriculture, Commerce, and Labor Departments was to review all classified and otherwise withheld sci-tech information that had been or would be generated with federal funds for the purpose of determining whether it could be released to the public. John W. Snyder, director of War Mobilization and Conversion, was appointed chairman and director. Within two months the responsibility of the board was broadened by Executive Order 9604 to include enemy scientific and industrial information. The order stated that the government's policy was to be ". . . prompt, public, and free and general dissemination of enemy scientific and technical information." These two orders placed a heavy burden on the Publications Board. During the same year, the Commerce Department, by Departmental Order No. 5, established an Office of Declassification and Technical Service with John C. Green as director. This office was to be the operating arm for the Publications Board, and its functions, as stated by departmental order, were to ". . . serve as a clearinghouse for collecting, editing, and publishing of such scientific and technical data for the purpose of promoting economic expansion and development." The office concentrated first on captured foreign material that was forwarded to it by the War and Navy Departments, British Intelligence, the Office of Scientific Research and Development, and other government agencies.

As early as January 1946, U. S. civilian experts were in Europe selecting material for the Publications Board at the rate of 5,000 items a month. In addition, hundreds of military experts were examining and forwarding material to military collection centers in London, England, and Dayton, Ohio. Many of the duplicates from this program were sent to the Publications Board. The Office of Declassification and Technical Services abstracted and indexed the material and then, depending on the subject matter, forwarded the documents to the Army Medical Library, the Department of Agriculture Library, or the Library of Congress. These libraries were to reproduce the documents requested by organizations who had been alerted by the "Bibliography of Scientific and Technical Reports" which was published monthly by the Publications Board. The collections programs in Europe and later in the Far East quickly overloaded the Office of Declassification and Technical Services, as well as the three libraries that housed and reproduced the material upon request.

In July 1946 Commerce Department Order No. 5 was amended again. The office's name was changed to the Office of Technical Services (OTS) and its functions expanded to include ". . . aid in increasing the technical production of the nation by assisting manufactures and industries by contacting federal, state, and private agencies which can help . . . by furnishing solutions to technical problems." OTS was placed under the assistant secretary of foreign and domestic commerce. By the end of 1947 it had received about five million documents from Germany alone. Many were on 35-mm film, which presented document reproduction problems for OTS and the cooperating libraries. The hugh task of reproducing legible copy in large quantities and the difficulty of achieving good legibility in so many documents added to the overload problem. In 1948 OTS received congressional approval to establish a revolving fund using the receipts of sales of copies of documents and bibliographies to support its program. However, during this year the acquisition of foreign material almost ceased, and OTS's federal funding and sales receipts were reduced. Even though the flow of foreign documents almost ceased, OTS's work load remained quite heavy, and its reduced income forced OTS to slow its production schedule, which caused customer complaints about the slow delivery of orders.

On September 9, 1950 Congress passed PL 81-777, which authorized and directed the Secretary of Commerce

> ... to establish and maintain within the Department of Commerce a clearinghouse for the collection and dissemination of scientific, technical, and engineering information, and to this end to take such steps as he may deem necessary and desirable
> (a) to search for, collect, classify, coordinate, integrate, record and catalog such information from whatever sources, foreign and domestic that may be available;
> (b) to make such information available to industry and business, to state and local governments, to other agencies of the federal government, and to the general public, through the preparation of abstracts, digests, translations, bibliographies, indexes, and microfilm and other reproductions, for distribution either directly of by utilization of business, trade, technical, and scientific publications and services;
> (c) to effect within the limits of his authority as now or hereafter defined by law and with the consent of competent authority, the removal of restrictions on the dissemination of scientific and technical data in cases where consideration of national security permit the release of such data for the benefit of industry and business.

The Secretary of Commerce assigned the above responsibilities to OTS. Thus OTS had a broad mandate to make both foreign and domestic sci-tech information available to the U. S. scientific, technical, and general publics.

However, OTS was beset with a number of problems. The first, reduced income, was mentioned previously. Another problem was the lack of control of reproduction and distribution of technical reports. These operations were assigned to other units of the Department of Commerce whose work schedules were not responsive to OTS's need for prompt delivery of requested documents. Further, questions were raised as to the uses that could be made of its revolving fund in which income from sales were placed. So in 1954 the Commerce Department asked the comptroller general for an interpretation of PL 81-776 on whether the department had (a) authority to prepare technical reports, (b) authority to use revolving funds to acquire or prepare technical reports, and (c) the right to charge a private party the total costs related to the preparation and distribution of a technical report. The comptroller general answered affirmatively on (a) and negatively on (b); regarding (c) he ruled that only costs directly attributable to the private party's request could be so levied. Evidence in the record of the OTS appropriations hearings indicates that the House Appropriations Committee was suspicious that OTS would use the revolving funds to expand its operations beyond the limited program Congress had in mind.

Under the above restrictions OTS was limited to a program of acquisitions based on voluntary submission of reports by other agencies and organizations; a restricted effort in cataloging and indexing; and preparation of technical reviews and reports with funds transferred from other agencies. Under these circumstances, OTS was never able to approach the potential of the program envisioned in PL 81-776, which later resulted in Congress and the executive branch establishing other programs to fulfill part of the responsibility authorized in the act. In addition, two other important conditions affected the ability of OTS to fulfill this potential. First, Director John C. Green, who was primarily interested in the utilization of new technology, did not feel that the collection, reproduction, and distribution of documents had the highest priority; therefore, he consented to the distribution and reproduction functions being performed in other units of the Commerce Department. Second, the Commerce Department, after its initial support for larger congressional appropriations, allowed OTS to operate with insufficient funds.

From the time of its congressional authorization OTS emphasized service to small firms. It developed a series of services that were widely used, including the "Monthly Bibliography of Technical Reports," "Technical News Letters," press releases on new reports considered of immediate interest, and reproduction of technical reports of wide interest. In 1951 OTS expanded its effort to have unclassified technical reports from other agencies sent to its few federal agencies, and some universities and private foundations added OTS to their distribution lists for part of their output. However, the other agencies continued to maintain their own report dissemination programs. During this same year, OTS, under contract with the Economic Cooperation Administration (ECA), was able to extend its technical information assistance to foreign areas. At first this service was limited to the Marshall Plan countries in Europe, but ECA or its successor agencies broadened the contract so that under the Agency for International Development (AID), OTS was able to give technical information assistance to less-developed countries in many parts of the world. In 1958, by interagency agreement, OTS became the federal government's central point for the distribution of unclassified documents not of potential use to unfriendly countries. In 1959 OTS was recognized as the outlet for the unclassified technical reports of all federal agencies. Many agencies, however, still maintained their own initial distribution programs. Also in 1958, OTS and the Special Libraries

Association (SLA) agreed to issue a consolidated *"Monthly List of Translations"* to be published by OTS. SLA concentrated on acquiring and preparing entries for translations made by nonfederal organizations, while OTS was responsible for those prepared in the federal government. Eight translations depository libraries were established in 1959.

In a later chapter, changes in the role and support of OTS, which are reflected in the change of its name to the Clearing House for Federal Scientific and Technical Information (CFSTI) and later to the National Technical Information Service (NTIS), are discussed.

MEDICAL SCIENCES INFORMATION EXCHANGE[4]

The rapid growth and dispersion of R&D activities during and following World War II led to the initiation of a new type of information activity, namely, a system for maintaining a current inventory of on-going R&D projects. Prior to this period, most scientists and engineers were able to keep track of significant R&D in their respective areas of interest through meetings, journals, and personal contacts. However, the rapid growth of science and technology and the secrecy surrounding many projects caused R&D administrators, as well as scientists and engineers, to become concerned over needless duplication of effort. Administrators of multidiscipline research were the first to develop current inventories of on-going research and advanced development projects. By 1946 the Agriculture Research Service, the Public Health Service, and some of the military R&D agencies had initiated such inventories.

In fact, in 1946 the National Institutes of Health (NIH) sponsored a meeting of private and governmental R&D administrators in New York City to consider this problem. Following this conference a number of organizations began to furnish information to the research project inventory of the NIH that covered the medical sciences. That same year the U. S. Interdepartmental Committee for Research and Development held a series of meetings to consider establishing a government-wide inventory. The outcome of these conferences was agreement to set up such an inventory in the medical sciences. For several legal and administrative reasons NIH did not wish to aceept the responsibility for this inventory, and it was decided to place the project under the National Research Council of the National Academy of Sciences (NAS/NRC).

The medical Sciences Information Exchange (MSIE), as this inventory was called, began operation July 1, 1950; it was funded through an Atomic Energy Commission contract for $60,000. Other agencies transferred part of their funds to the AEC for use in support of the MSIE, which was placed under the administration of NRC's Medical Sciences Division, headed by Keith Cannon. Policy and program planning for the project was carried out by a committee of representatives of NIH, the offices of the surgeon generals of the Army and Air

Force, the Office of Naval Research, the Atomic Energy Commission, the Veterans' Administration, and the NRC. Stella Deignan, who had been director of the NIH Office of Exchange of Information, was transferred to NRC and became Director of MSIE.

The following excerpts from the agreement signed by the participating parties best describes the goals of MSIE and its cooperating agencies.[5]

Scope
a. To limit the activities of the Medical Sciences Information Exchange to the accumulation, organization, analysis, and distribution of information concerned with current research in medical and allied fields. Information may not be released in any form or to any person not authorized under the charter, or defined below except with the approval of the agency or agencies concerned.

Information to be furnished to the MSIE
b. To supply information to the Medical Sciences Information Exchange by utilizing (in connection with applications for research support or early contract negotiations) the "Notice of Research Project" in use in the present Office of Exchange of Information.

Information to be routinely available from MSIE
c. To authorize the Medical Sciences Information Exchange to make routinely available to cooperating agencies and authorized groups and individuals the following types of information:
   1. Amounts and sources of support of geographical areas, research institutions, departments, and investigators;
   2. Amounts and sources of support for the board and specific areas of research listed in the subject index being developed in the Medical Sciences Information Exchange;
   3. Lists of investigators and institutions engaged in special types of research.

Types of information not to be available from the MSIE
d. To instruct the Director of the Medical Sciences Information Exchange:
   1. To withhold from any organization, cooperating or otherwise, information in any form for the purpose of enabling a determination of the programmatic, fiscal, or scientific program of any agency;
   2. To refrain from supplying evaluations of individual projects or programs of one agency to another;
   3. To decline requests to supply information in any manner not routinely available in the office when the compliance with such requests would place an undue burden upon the staff.

Operation of the Exchange
e. To delegate operation of the activity to a Director who shall have the authority to determine his own needs for staff, space, and equipment within the limits of the recommended budget and the policies agreed upon by the National Research Council and the Policy Committee of the Medical Sciences Information Exchange. The Director shall be authorized to exercise judgment in responding to requests, and to call meetings of the Policy Committee when changes in or divergence from policy are involved or to clear with the pertinent agency when a single agency is involved.
It is agreed that no cooperating agency may interfere in the internal operation of the activity.

Science Foundation
e. That the status of the Medical Sciences Information Exchange will be reviewed after the Science Foundation has begun to operate to determine the most appropriate location and administrative arrangements for the exchange.

During the next two years MSIE became a useful center for information ongoing extramural research projects.

BIOLOGICAL SCIENCES INFORMATION EXCHANGE[6]

Several agencies had been urging for some time that the topical scope of the Medical Sciences Information Exchange be expanded and that intramural research projects also be included, and in March 1953 the Policy Committee presented the National Research Council with an agreement that had been approved by the participating agencies. The significant changes from the 1950 pattern were:

1. MSIE's name was changed from Medical Sciences Information Exchange to the Biological Sciences Information Exchange (BSIE). Biological sciences were interpreted to include the medical sciences, anthropology, psychology, and some areas of the physical sciences of interest to the Agriculture Department.
2. The funding pattern was changed to allow each agency to transfer funds to the NRC, but no one agency could finance more than 25 percent of the total basic costs of BSIE. An informal agreement among the federal agencies allowed an agency to pay for special services from BSIE.
3. Nonfederal organizations could participate in the exchange. This provision allowed private foundations, state agencies, and private research institutes (including universities) to register their research projects and request information from BSIE.

4. NRC representation on the BSIE Policy Committee was expanded to include divisions of anthropology and psychology as well as biology and agriculture.
5. The director of the exchange could issue an annual report if he wished, but it was subject to the approval of the Policy Committee.
6. The director of the exchange could routinely disseminate information in the exchange's field as long as the scope, size, and character of an individual agency's research program were not revealed. Federal agency representatives did not want BSIE furnishing such information to other federal agencies, to the Bureau of the Budget, or to congressional committees.

The National Research Council anticipated difficulties in administering the broadened and enlarged program outline for the exchange. Following consideration of alternate organizational locations, Leonard Carmaichel, secretary of the Smithsonian Institution, and Ernest H. Allen, chairman of the Policy Committee, reached an agreement for the transfer of the BSIE to the Smithsonian Institution under the terms of the agreement submitted to the NRC. The participating agencies accepted this proposal, and BSIE began operations under the Smithsonian in July 1954. BSIE continued to enlarge its operation during the next five years as the research programs of the federal agencies grew. During this period many suggestions were made to include the physical sciences in the inventory program. Finally in 1960, with the concurrence of the Bureau of the Budget and the Federal Council for Science and Technology, BSIE was so expanded. This enlarged role is reviewed in Chapter 4.

LIBRARY OF CONGRESS[7]

During World War II the Library of Congress made no significant policy or program changes in its activities related to science and technology. However, the collections of earth, environmental, and social sciences were heavily used by the agencies involved in planning and conducting the war, and this intensive use identified many critical gaps in the Library's collections in these fields. These deficiencies were especially evident in foreign materials. Areas of particular concern were geology; geography; regional studies; census, and statistical data on materials, location, and character of industries; and distribution and composition of populations of foreign countries. As soon as channels were open into the war-affected areas, the Library increased its acquisition program, including by sending individuals and teams on procurement missions to Europe, Asia, and South America.

Beginning in 1947, a series of contractual agreements with other federal agencies enabled the Library of Congress to expand its science and technology programs. As mentioned earlier, in 1947 the Library accepted funds from the Office of Naval Research to perform numerous tasks and established the Science and Technology Project (STP) to fulfill the conditions of its contract with ONR. Although STP's services were intended primarily for the benefit of the Navy, ONR soon gave permission for the Library to make its publications and services available to other federal agencies and their R&D contractors. During 1949 the project was renamed the Navy Research Section (NRS), and another important addition to its activities was the preparation of special bibliographies for federal agencies, which transferred funds for this purpose. Also in 1949 the Library accepted funds from the Air Force for another division to document and prepare bibliographies, abstracts, and reports on special topics. This project worked largely in scientific and technical fields, with the major interest in exploiting material from the Eastern European bloc. Finally, near the end of 1949, the Library accepted a contract from the Snow, Ice, and Permafrost Establishment (SIPRE) of the U. S. Army Corps of Engineers to identify, abstract, and index reports and other publications of interest in SIPRE's research programs. These abstracts were published in both card and bulletin form and were distributed to individuals and organizations authorized by SIPRE.

The pressure on the Library of Congress from these projects as well as from students and scholars in science and engineering stimulated the Library to establish a Science Division in 1949 with Raymund Zwemer as chief. Its major purpose was to improve the acquisitions of journals, serials, and monographs as well as to furnish more responsive reference and bibliographic services. The Science Division had only nominal technical and administrative relationship to the Air Force and Navy projects but was directly involved in the administration of the SIPRE project and the Aeronautics Section, which had been established with Guggenheim funds in 1929. In 1952 a Technical Information Division, in addition to the Science Division, was formed and consisted of the SIPRE Project and the Navy Research Section. In 1953, however, the Armed Services Technical Information Agency (ASTIA) assumed policy responsibility for the Navy Research Section (renamed ASTIA Reference Center), except for the bibliographic unit, which performed bibliographic services for agencies other than ASTIA. During the same year the Air Force projects that were exploiting the Eastern European literature were greatly expanded and raised to divisional status. For the next five years, the Library of Congress science and technology programs continued along the lines indicated above until, in 1958, two organizational changes were made: First, the ASTIA Reference Center was moved from the Library and consolidated with two other ASTIA branches in Arlington Hall Station in Virginia, and second, the Science Division and the Technical Information Division were united into a Science and Technology Division.

In sum, the Library of Congress science and technology activities changed during a ten-year period from a passive program of acquisition, cataloging, and limited service to an active and much enlarged program responding to the special bibliographic and other needs of the executive agencies as well as those of the scholarly community and the general public.

TECHNICAL INFORMATION DIVISION, ATOMIC ENERGY COMMISSION[8]

Scientific and technical information activities in the field of atomic and nuclear energy considerably predate the establishment of the Atomic Energy Commission (AEC). Early in its existence, the predecessor agency to AEC, the Manhattan District Project, organized libraries and other scientific and technical information activities within each of its laboratories. Because of the secrecy surrounding the activities of each Manhattan District Project laboratory during World War II, preparation of scientific and technical reports, special bibliographies, and catalogs was tightly compartmentalized. By 1945, when hostilities had ceased, the scientists and administrators of the project became concerned over the lack of effective communications among laboratory scientists and engineers, as well as the duplication of effort in library, bibliographic, and acquisitions activities. During this same year, the project administration became convinced that means had to be devised to evaluate, organize, and publish the record of scientific and engineering advances accomplished in the laboratories. Therefore, a Manhattan District Editorial Advisory Board (MDEAB) was organized. This board had at least one representative from each of the laboratories, including Alberto F. Thompson from Oak Ridge and Herman H. Fussler from Chicago. The responsibility assigned to the MDEAB was to plan a series of monographs that would summarize the state of art in the various fields of nuclear energy. These monographs became known as the "National Nuclear Energy Series." In addition to planning and preparing this series, MDEAB investigated ways in which library and technical information activities of the project could be made more effective. Early in 1946 Herman Fussler prepared a memorandum outlining his suggestions for a scientific and technical information service in nuclear energy. Among his recommendations were the following:

1. As complete a set as possible of all scientific and technical reports on nuclear energy, including British and Canadian documents, should be established in the Research Section at Oak Ridge.
2. Contractors should be required to submit project progress and final reports to the Research Section.
3. Subject categories should be established for reports.
4. The Research Section in Oak Ridge should be capable of reproducing good, legible copy of manuscripts that were submitted.
5. A method of distributing reports, utilizing subject categories as a basis, should be established.

6. The Research Section should do detailed subject indexing of reports by utilizing IBM punchcards.
7. The Research Section should prepare an abstract journal in which research reports would be categorized under broad subject headings.
8. The abstract journal also should contain abstracts of scientific journal articles, especially those of foreign origin.

It was several years before all of Fussler's recommendations were implemented, but they did provide guidelines for the development of AEC's technical information program.

By 1944, the Manhattan District Project had developed a large, effective sci-tech information service. In the spring of 1946, the Library Unit of the Research Division in the Oak Ridge National Laboratory was organized with Bernard M. Fry as chief. The unit was assigned the following responsibilities for the entire Manhattan District Project: control and exchange sci-tech material among the units of the Manhattan District Project; arrange for review of certain types of material for early release to the public; and gather and review all sci-tech material produced by the project with a view to declassification. By April 1946 this unit, with aid of a project-wide committee, issued a declassification guide that was distributed to all project groups and that did much to standardize the further declassification of project materials. Soon thereafter a Bibliography and Literature Survey Section was organized in the library to alert project personnel to material being received. The Editorial Advisory Board recommended issuance of a "Weekly Title List," which the library began in June 1946. This publication was the forerunner of the "Nuclear Science Abstracts" that became a major bibliographic aid to nuclear research and engineering.

The problems of organizing comprehensive subject and other indexes immediately faced the new project library. In late 1946 and early 1947 after passage of the AEC enabling act, consideration was given to the use and expansion of the index of materials received by the Metallurgical Laboratory in Chicago; this index, which was being prepared by Yolande D. Young, provided the base for the AEC catalog. During the same period the list of subject headings used by the library of the Chicago laboratory was expanded and modified for use by all project information services. Another effective dissemination tool in the nuclear energy field was the classified distribution guide that was issued to all units and contractors of the new AEC in April 1947. The procedures outlined in this guide enabled the AEC to control the automatic distribution of reports by its laboratories and contractors, while at the same time allowing for early distribution by the issuing units.

The Atomic Energy Act, as passed in October 1946, had included the following provision: "The dissemination of scientific and technical information relating to atomic energy shall be permitted and encouraged so as to provide that free interchange of ideas and criticism which is essential to scientific and industrial progress and public understanding and to enlarge the fund of information." The commissioners of AEC took this section of the enabling act seriously and in establishing the Division of Public and Technical Information in October 1947, stated that the division was ". . . to secure, in accordance with the Atomic Energy Act, adequate control and dissemination of information that is the life blood of scientific and engineering progress—and of public understanding of that progress and its implications."

To implement the above directive, the AEC assigned the following functions to the division:

1. Preparation, reproduction, and controlled distribution within the AEC of classified reports to personnel requiring such reports;
2. Declassification, through analysis by qualified scientific reviewers, of material that could be made available for public use;
3. Editing, preparation, and reproduction of technical information material, including indices, abstracts, project reports, and declassified papers;
4. Security guidance to aid publishers, editors, reporters, broadcasters, leaders of citizens groups, and others issuing material relating to atomic energy in preventing compromise of restricted data and to assist AEC contractors and other agencies of the government in safeguarding such information.
5. Provision of public information service, including assisting representatives of the press, radio, picture services, citizen and trade organizations, and educational agencies, to obtain the full range of declassified and unclassified data currently available from the national atomic energy program.

The task of organizing an effective national information service was enormous even though the basic plans and organization had been developed during the last years of the Manhattan District Project. While the details of the development of the AEC sci-tech information program cannot all be included here, the following text identifies the major decisions and actions that resulted in the AEC's developing one of the world's most effective information services through the Division of Public and Technical Information.

Morse Salisbury, formerly director of information in the Department of Agriculture and director of public information for the UN Relief and Rehabilitation Administration, was appointed head of the new division. He established an office in the Washington, D. C., headquarters and organized the division into three branches: Declassification Branch, Public Information Branch, and Technical Information Branch whose operations were located in Oak Ridge. Alberto Thompson, who had directed the technical information activities at the Oak Ridge National Laboratory, was named chief of the Technical Information Branch. He had been brought to Washington earlier to assist the AEC in planning its sci-tech information program. Brewer F. Boardman assumed A. Thompson's duties in Oak Ridge. In March 1948 the Manhattan District Editorial Advisory Board was replaced by the Technical Information Panel. Both the groups included individuals intimately involved in the research and engineering activities of the various AEC and contractor laboratories. As was the case with the editorial board, the new panel played an important role in influencing the character of the development of the Technical Information Branch.

Very early in the Technical Information Branch's history it became apparent that it was not receiving all the scientific and technical reports of the laboratories and contractors of AEC, even though this branch had responsibility for centralized processing, reproduction, and dissemination. In April 1948 Bulletin CM-81 was issued and gave specific guidelines for control of report literature. This bulletin emphasized that all project sites must send one copy of every report to the Technical Information Branch at Oak Ridge. Close observance and careful monitoring of this directive did much to give the Technical Information Branch effective control of all AEC technical report literature. Another effective mechanism instituted by the Technical Information Branch was a series of conferences of all AEC librarians, including those from its major contractors. At the first meeting, in April 1948, the group recommended that copies of all translations of foreign nuclear science publications be forwarded to Oak Ridge and that an indexed current list of these be issued. The recommendation was accepted and the first consolidated list and subject index was issued in January 1949 under the title of *Library Bulletin*. This bulletin included a list of special bibliographies prepared in the AEC laboratories. Late in 1948 the Technical Information Branch at Oak Ridge was placed under the administrative and funding control of the Division of Public and Technical Information in Washington, D. C. Thus a key operation of the division was removed from a laboratory and placed under the national headquarters, thereby allowing for more effective operational and administrative control.

In December 1948, the first fruits of the program designed by the Manhattan District Advisory Editorial Board became available; this was the first volume of the "National Nuclear Energy Series." The administrative and fiscal problems raised by this endeavor are worthy of note. AEC had persuaded the Joint Congressional Committee on Printing to grant a waiver from the requirement for printing by the Government Printing Office. The waiver was given with the condition that no federal funds would be used for printing and binding. Next, the AEC contracted with Columbia University for the preparation of the manuscripts, and McGraw-Hill Book Company was given a contract for publication and sale, which included a stated sales price with a specified royalty. AEC was to receive copies of the volumes in lieu of royalty, up to a certain amount. Royalties above that predetermined figure were to be paid to the U. S. Treasury. (These fiscal and administrative guidelines were used subsequently by other agencies to produce high-quality sci-tech publications—that is, by using scientists outside the government and taking advantage of the competence of the commercial publishers.) Most of the administrative, technical, and fiscal responsibilities for the nuclear energy series rested with the Technical Information Branch, but the research divisions of the laboratories participated in the selection of authors and topics and reviewed the manuscript for quality.

Among the Technical Information Branch's first problems were the production of catalog cards and the reproduction of reports and other materials for use in the laboratories. To be of maximum usefulness to the scientists and engineers, these operations had to be performed on a tight schedule. AEC petitioned the Joint Congressional Committee on Printing for a field printing plant at Oak Ridge. This request was granted, and the printing plant was placed under the direction of the Technical Information Branch, which allowed it to determine priorities and, even more importantly, enabled branch personnel to experiment with innovations that did much to expedite and improve the quality of the AEC's information products. The Joint Committee also approved sales of these materials by AEC. Early in 1949 AEC transferred its sales operations to the Department of Commerce's Office of Technical Services; however, OTS proved unable to handle the workload so this task was returned to Oak Ridge. In 1950 when an executive order was issued giving OTS authority to sell all government publications not printed by the Government Printing Office, AEC again returned the sales operations to OTS and concentrated on automatic and request distribution of its materials. This pattern led to an effective distribution program, which in later years placed a heavy financial burden on the AEC. In 1950 the AEC responded to a suggestion by E. E. Thum of the American Society of Metals by establishing thirty-one regional depositories in libraries designated by a committee of the American Library Association. These libraries were furnished the necessary reports and bibliographic publications to enable them to give service on the AEC documents; a few similar arrangements were made with foreign centers.

Near the end of 1950, the Technical Information Branch's activities had expanded to a size such that the AEC raised the branch's administrative status and renamed the activity the Technical Information Service (TIS); the Oak Ridge technical information operation was placed under its administration. TIS still reported to the Division of Public and Technical Information but had considerable freedom for administrative, fiscal, and technical decisions. Throughout its history, TIS adjusted its services to meet the changing needs of scientists, engineers, and laboratory librarians. Among its products were the *Weekly Title List*, which early included abstracts and indexes and which was eventually renamed *Nuclear Science Abstracts* (NSA). The *Library Bulletin* was issued in response to the need for information on translations and specialized bibliographies. In 1951, another service was introduced that deviated from the usual production routine. At a conference of AEC librarians at Sandia Laboratory in New Mexico, plans were made for a specialized service on weapons development.

Prior to this time, cataloging and indexing in this field had been done at the Oak Ridge center. At this Sandia meeting, agreement was reached on principles and rules for cataloging, indexing, and filing so that compatible cataloging and indexing of reports containing weapons information could be carried out at the various projects. This system was put into effect, with data from the several sites being forwarded to Oak Ridge for review, reproduction, and distribution. Experience gained in this project laid the basis for later decentralized projects.

Technological innovations for production of catalogs, indexes, bibliographies, reports, and other information materials were an important part of the development of the AEC technical information program. For example, as early as 1948 the Oak Ridge center began to produce the abstract bulletin and index and title lists by shingle layering of catalog cards in page format so photo-offset could be used for printing the publications. In 1952, when laboratories and other AEC centers no longer had space to store back issues of technical publications, the TIS contracted with the Microcard Corporation to produce microcards of older reports. After a few months, because of the usefulness and clarity of the early microcards, all current reports were reproduced in microcard form, which thus became part of the standard distribution and enabled many more centers to hold comprehensive collections of AEC reports, as well as the abstract bulletins, indexes, and special bibliographies. Careful attention to the design of the microcard format, high-quality image, and capability for reproduction in original page size were all essential elements in the success of this service. Maintaining the microcard production laboratory on the premises at Oak Ridge enabled AEC to meet reasonable production schedules.[9] In 1958, after considerable experimentation under the direction of Melvin Day and Robert Shannon, the TIS extension in Oak Ridge developed a mechanized system for preparing indexes using the Listomatic camera and IBM punchcard equipment. The first indexes prepared in this manner were included in the issue of *Nuclear Science Abstracts* for January 15, 1959. This new procedure allowed for inclusion of a current indexes in each issue of *NSA*. By January 1960, this operation system had been improved so that many indexes and bibliographies could be prepared automatically for printing. Thus, AEC was able to reduce still further the time lag in the issuance of many of its bibliographic tools.

In response to the concern that innovations and new scientific findings in nuclear science were not being utilized in industrial and other sectors of American life, an Advisory Committee on Technological Information was formed in 1952. The members were largely representative of the scientific, technical, and industrial press. These editors and publishers were briefed on new findings and encouraged to publish the information in their journals and news magazines. Although this experiment met with some success, in 1954 AEC established an Industrial Information Branch in TIS under the administration of Edward J. Brunenkant. This branch's assignment was to work closely with industrial firms to acquaint them with developments in the atomic energy field that they might find useful. This step was made possible by the 1954 Atomic Energy Act that authorized AEC to make more information available to U. S. industry as well as to foreign countries. This action was stimulated by the Atoms-for-Peace program of 1953 that was launched by President Eisenhower.

Following passage of the 1954 act, AEC made available to foreign centers a basic collection of documents on atomic energy, accompanied by the pertinent catalogs, abstract bulletins, indexes, and bibliographies. By 1955, AEC had concluded agreements for cooperative exchange of information with the United Kingdom, Canada, Belgium, Switzerland, and the Netherlands. These were the first steps in a series that ultimately led to cooperation with EURATOM, a consortium of European countries for research and development in atomic energy. Later came the initiation of the International Nuclear Information Service (INIS) under the sponsorship of the International Atomic Energy Agency (IAEA). In 1957, TIS was authorized to exchange publications with Eastern bloc countries. Although these exchange agreements did enable TIS to acquire some useful scientific material, for a variety of reasons this aspect of the program never was considered very successful. One reason was that the return flow was not very great nor was the content of the publications very useful in determining the characteristics and direction of the participating countries' R&D.

In sum, the AEC/TIS is an example of a well-planned, -organized, and -operated federal information service. Several factors contributed to the success of this service: first, the AEC commissioners recognized the importance of this activity to the furtherance of the agency's mission and therefore provided an administrative and fiscal climate that was responsive to the needs of the service. Second, the AEC mission was clearly defined, and it had the necessary authority to control nuclear information in the U. S. Third, the people who planned and operated the service were dedicated, innovative, and competent. Discussion in Chapters 4 and 5 will show that the AEC information services continued to expand and improve; more importantly, the techinques, services, and types of products were adopted or adapted by other U. S. and foreign information services.

OTHER FEDERAL AGENCIES

During the same time span described in the foregoing discussion on the development of new federal information organization and services, many established federal library and other information centers experienced marked changes in their programs and operations. The following discussion briefly characterizes some of these changes.

Army Medical Library

In 1948, a Department of Defense study committee recommended that the Army Medical Library become a joint agency of the three military services. It was four years, however, before the Secretary of Defense issued the order changing the administrative status of the Library to the Armed Forces Medical Library.[10] In 1955, a Hoover Commission Task Force recommended that the Armed Forces Medical Library be recognized as the National Library of Medicine. This recommendation was followed by congressional action that resulted in the National Library of Medicine Act (PL-941, 84th Cong. 70 Stat. 960) in 1956. During the consideration of this legislation, the need for and the location of a new building was given favorable review by both the Senate and House committees; however, these committees recommended in the conference report that the new National Library of Medicine's Board of Regents make these decisions.

The library's program was greatly expanded during the 1942-1956 interval, with a great increase in collections and significant changes in services. In 1943, the library implemented a new policy whereby government agencies and individuals connected with accredited institutions could receive free microfilm copies of material in lieu of loan of the original. (This policy was modified in 1957 to require that all individual loan requests be channelled through a library and that interlibrary loan and photo duplication services be considered two phases of the same plan, with the library furnishing photocopies in lieu of the original work if costs, legal restrictions, and other pertinent conditions warranted.) In 1950 the library issued its first annual catalog, *Army Medical Library Author Catalog*, including a subject index, which replaced the *Index Catalogue* that had been started in 1880. The same year the library awarded a contract to the Welsh Medical Library of Johns Hopkins University for a project to study problems of medical bibliography with emphasis on applications of machine methods. The findings of the project, headed by Sanford Larkey, encouraged the library to continue its investigation of automated techniques and finally led to the initiation of the MEDLARS development program, which is discussed in later chapters.

Department of Agriculture Library[11]

In 1942 the Department of Agriculture completed a series of fiscal and administrative moves that placed all departmental libraries under the fiscal and program control of a director of libraries, who at that time was Ralph Shaw. In the same year the Department of Agriculture Library issued the first volume of the *Bibliography of Agriculture*. In 1950 the library initiated the practice of contracting for library services for departmental employees where no departmental library service was available. In 1952, the microfilm system, known as the "rapid selector," was introduced into the acquisitions and cataloging operations of the library. The rapid selector scanned coded bibliographic references on microfilm, and when a selected code was identified, it would reproduce a copy of the reference. In 1959, the Department of Agriculture again decentralized the funding and administrative management of all departmental libraries outside the metropolitan area of Washington, D. C. These responsibilities were placed with the agencies that the libraries served. The director of libraries continued to act as a coordinator of these spacial and regional libraries.

Department of Interior Libraries

The collections and services of the Department of Interior's bureau, field, and departmental libraries experience the same accelerated growth as other major sci-tech services from 1942 to 1956.[12] Near the end of this period, the Department initiated a series of studies that later led to coordination of these libraries under a director of libraries. Also several bureau libraries were incorporated into the departmental library.[13]

Military Sci-Tech Information Services

The military sci-tech information services were greatly expanded and many new units were established. Large centralized photographic collections were organized by the Navy and the Air Force that included selected aerial photographs of combat areas. Library and other sci-tech information services were expanded and mechanized operations were introduced in the specialized military centers such as the Navy Hydrographic Office in Suitland, Maryland; the Army's environmental research establishment at Natick, Massachusetts; the Air Force's Technical Intelligence Center at Dayton, Ohio; the Army's Map Center near Bethesda, Maryland; and the Air Force's Aeronautical Chart Center near St. Louis, Missouri.

The Defense Department, through its contracts with large corporations, enabled these organizations to expand, disperse, and automate their technical information services. Among these were the information services of United Aircraft Corporation on Long Island; the General Electric establishment at Schenectady, New York, the General Motors Research Center near Detroit, Michigan; and the Dupont Corporation's facilities in the Wilmington, Delaware, area.

CONCLUSION

The 1942-1956 era can be characterized as the period when many large federal sci-tech information services were organized and other federal sci-tech libraries and specialized information centers were greatly expanded. The era was also one of remarkable increases in data collections related to nuclear, environmental, physical, and chemical sciences. By the end of this period, federal as well as nonfederal libraries and information centers were striving to develop intellectual, administrative, and operational mechanisms that could cope with the expanding flow of information in many forms as well as the increasing numbers, types, and varieties of service demands by the growing population of scientists and engineers. Pervading the thinking of the producers, organizers, and distributors of the growing corpus of scientific information was the realization that the rapidly developing new technologies in photography, reprography, computers and other electronic devices, and telecommunications should be effectively utilized to upgrade and expedite the storage, handling, and use of sci-tech information.

NOTES AND REFERENCES

1. Sydney Passman, *Task Group Report on the Role of the Technical Report in Scientific and Technical Communication* (Springfield, Va.: Clearinghouse for Federal Scientific and Technical Information, 1968), 112 pp.; and Bernard Fry, *Library Organization and Management of Technical Report Literature* (Washington, D. C.: Catholic University of America Press, 1953), 140 pp. The latter is the first detailed discussion of the problems of organizing and servicing sci-tech reports and was originally submitted as a requirement for a master's degree in library science.
2. See Defense Documentation Center, *DDC: Origins and Milestones* (Alexander, Va.: Defense Documentation Center, 1971), 13 pp. Pages 1-3 identify important developments of DDC's predecessor agencies to 1956.
3. See John D. Gallivan, III, "A History of the Clearinghouse for Federal Scientific and Technical Information," term paper for Seminar in Management Information and Operations Control, American University, Washington, D. C., 124 pp. Pages 1-57 give data on the Office of Technical Services to 1957.
4. See Smithsonian Science Information Exchange, records on the history of the exchange. The MSIE contract was signed by representatives of the National Research Council, U. S. Atomic Energy Commission, Veteran's Administration, Surgeon General's Office of the Army, Office of Naval Research, Public Health Service, and the Surgeon General's Office of the Air Force.

5. Smithsonian Science Information Exchange. Data on the Biological Information Exchange is from the records of SSIE. Contracting agencies were AEC, Surgeon General's Office of the Army and Air Force, Public Health Service, and the Smithsonian Institution. The National Science Foundation was represented but withheld support until its program was better formulated.
6. See ibid.
7. Library of Congress, *Science and Technology in the Library of Congress* (Washington, D. C.: Science and Technology Division, Reference Department, Library of Congress, 1965), 10 pp., provides a brief chronology.
8. See Robert B. Devine, *Milestone in the Evolution of an Information System: Development of the U. S. AEC's Information Program* (paper submitted to the Division of Technical Information, AEC 1961?), gives an account of the early development of AEC's technical information activities. Also see R. L. Shannon, *Nuclear Science Abstracts: A 21-Year Perspective in Handling of Nuclear Information*, Document No. 1AEA-SM-128-32 (Vienna: International Atomic Energy Agency, 1972); and United States Atomic Energy Commission, minutes of the Technical Information Panel meetings, 1948 to 1972.
9. United States Atomic Energy Commission, *The AEC Program for Disseminating Technical Information*, TID-4551 (Oak Ridge, Tenn.: U. S. Atomic Energy Commission, Technical Information Division, 1953), 22 pp., describes in some detail AEC/TIS's products and services.
10. Manfred J. Waserman, "Historical Chronology and Selected Bibliography Relating to the National Library of Medicine," *Bulletin of Medical Library Association* 60, No. 4 (1972): 551-558. On pages 553-55, Waserman identifies significant actions relating to NLM during the period covered by this chapter.
11. See Foster E. Mohrhardt, "The Library of the United States Department of Agriculture," *Library Quarterly* 27, No. 2 (1957): 61-82. Pages 71-82 review activities of NAL during the period covered by this chapter.
12. United States Department of Interior, *Years of Progress 1945-1952* (Washington, D. C.: Government Printing Office, 1953), 195 pp., provides an account of Interior Department's programs. The discussion of sci-tech information activities are treated incidentally in relation to other programs.
13. See, for example, John C. Rabbit and Mary C. Rabbit, "The U. S. Geological Survey: 75 Years of Service to the Nation, 1879-1954," *Science* 119 (May 28, 1954): 741-58.

Chapter 4

The Decade of Investigation, Reorganization, and Expansion, 1957-1966

As indicated in Chapter 3, information services expanded rapidly as part of the marked growth of R&D during and following World War II; many libraries and specialized information centers were enlarged, and in reponse to the increased use of sci-tech reports, a number of present technical information centers were established. Near the end of the period, however, congressional committees and policy and program officers in the executive branch began to question the effectiveness of this pell-mell expansion. The uneasiness was reinforced during the early part of the 1957-1966 decade by a sense that the United States might be falling behind other nations in science and technology.

Of the foreign advances, the launching of Sputnik in 1957 had the greatest impact on Americans who had been led to believe that the United States was the leader in most fields of science and engineering. The knowledge that the National Science Foundation (NSF), in cooperation with the military agencies, was preparing to launch a satellite did not change the disconcerting facts that it had not done so or that U. S. space research specialists did not realize the USSR had developed strong capabilities in advanced rocketry and guidance systems. Two other foreign activities came to the attention of the government officials and added to the uneasiness at this time. First, information regarding a large sci-tech information center in the USSR began to circulate in the U. S. This center had been developing since 1952 and by 1957 had expanded into one of the largest such centers in the world. The All-Union Institute of Scientific and Technical Information—known as VINITI—had been established by the Soviets to collect, organize, and abstract the world's sci-tech information and disseminate it to Russian scientists and engineers. It also had developed a large translation center, and U. S. officials held the belief, which later proved to be erroneous, that this Soviet institute had a very advanced mechanized information system. Second, at about the same time, in 1957, Japan announced the establishment of a major R&D agency in which it had organized a large technical information center called the Japan Information Center for Science and Technology (JICST).

These and other incidents stimulated congressional committees and executive branch officials to ask the R&D community why these countries were outdistancing the United States in the information services field, and what this country should do about it. Without really waiting for the results of any investigations, Congress and the executive branch set out to improve the nation's scientific information capabilities. After considering several alternatives, in 1958 the president recommended and Congress passed the National Defense Education Act, Title IX of which authorized and directed the National Science Foundation to establish a Science Information Service with the following responsibilities: to provide or arrange for the provision of, indexing, abstracting, translating, and other services leading to a more effective dissemination of scientific information, and to undertake programs to develop new or improved methods, including mechanized systems, for making scientific information available. The capabilities of many existing libraries and specialized information centers were also strengthened; these included the National Library of Medicine, the technical information division of the National Advisory Committee on Aeronautics, and the military information services.

INVESTIGATIONS AT THE NATIONAL LEVEL

A number of investigations were initiated to identify the weaknesses in the sci-tech information services both inside and outside the federal government. The President's Science Advisory Committee (PSAC) appointed a panel, first chaired by Alan T. Waterman, director of the National Science Foundation, and later by William O. Baker of Bell Research Laboratories, formerly Bell Telephone Laboratories, to investigate the matter and make recommendations for improvement. This committee began calling witnesses from both the governmental and private sectors early in 1958. By the end of that year, the committee issued a report,[1] upon the basis of which the president asked the National Science Foundation to strengthen its science information program with the purpose of stimulating the improvement of scientific information services in the United States. The Baker report stressed that the United States should build on its present libraries and other information services, but not try to centralize them (see Appendix C).

This action was followed in March 1959 by an amendment to Executive Order 10521. The new Section 10 included this paragraph: "The National Science Foundation shall provide leadership in the effective coordination of scientific information activities of the federal government with a view to improving the availability and dissemination of scientific information. Federal agencies shall cooperate with and assist the National Science Foundation in the performance of this function, to the extent permitted by law."

The Decade of Investigation, Reorganization, and Expansion 1957-1966   55

PSAC continued to monitor the sci-tech information activities as did the Federal Council for Science and Technology (FCST) after it was established in 1961. However, these bodies and several presidential advisors for science and technology became dissatisfied with progress in improving sci-tech information programs. In 1961 Jerome Wiesner, then science advisor to the president, appointed a task force headed by James Crawford of the Oak Ridge National Laboratory. This group prepared an extensive report that included a number of recommendations for the improvement of federal information activities.[2] One of these was that each agency should designate an official who would have responsibility for coordinating the information activities of his agency or department. This recommendation was endorsed by both PSAC and the FCST, and several R&D agencies implemented the recommendations (see Appendix D for a summary of the recommendations).

Before the Crawford report was released, the PSAC felt that a broader look should be taken at sci-tech information activities, and it appointed a panel headed by Alvin Weinberg, director of the Oak Ridge National Laboratory, to review sci-tech information activities in both the federal and nonfederal sectors. This panel issued a report in 1963 that had wide dissemination and caused considerable interest both in the United States and abroad.[3] (It was translated into German and several other languages.) The panel's major recommendation was that it was essential for practicing scientists to be actively involved in scientific information programs if these services were to be of high quality and responsive to the requirements of the scientists. Also the report stressed the need for specialized information centers to evaluate information in the light of the needs of special groups. Weinberg took this recommendation seriously and established several new information services at the Oak Ridge National Laboratory. A few other government R&D agencies took similar action. In the early 1970s, as many as 117 information analysis centers were supported by federal agencies.

Congress also turned its attention to sci-tech information. The Senate Subcommittee on Reorganization and Internal Organization of the Senate Committee on Government Operations held a series of hearings during 1959 and 1960. This subcommittee, chaired by Senator Hubert H. Humphrey, issued a report entitled *Documentation, Indexing and Retrieval of Scientific Information.*[4] The Subcommittee did not draft legislation, but Senator Humphrey interrogated representatives from the private sector, heads of federal libraries and information centers, directors of R&D programs, and experts in the technologies related to library and information sciences. His major contribution was to insist that responsible government officials report back to the subcommittee on steps they were taking to improve the sci-tech information activities within their agencies and departments. Senator Humphrey's reporting requirement made many of the heads of R&D agencies aware for the first time of the problems in the information field and called to their attention some of the weaknesses in their agency's information activities.

In 1961 Senator John F. McCellan, chairman of the Senate Committee on

Government Operations, directed his committee's staff to investigate the R&D programs in the government. The report on this study was prepared by Edward Wenk of the Legislative Reference Service, Library of Congress. Part of Wenk's report was devoted to sci-tech information activities. Although no congressional action was taken as a result of these investigations, agencies recognized certain weaknesses in their information operations and took corrective actions. In addition to these special hearings, congressional authorization and appropriations committees increased their surveillance of the information activities in the agencies within their purview.

In 1964 the House of Representatives formed the Select Committee on Government Research, under the chairmanship of Congressman Carl Elliot. The committee's staff made an intensive study of the R&D programs in the federal government and subsequently issued a number of reports. One on the sci-tech information activities was titled *Documentation and Dissemination of Research and Development Results.*[5] While this report was the most comprehensive of the several congressional reports on sci-tech activities, Congressman Elliot was not re-elected to Congress, and the Select Committee did not pursue the sci-tech information activities further.

As a result of hearings held in 1963 and 1964, Congressman Roman C. Pucinski of Illinois conceived of a national center in Chicago with authority and funds to develop a national network of federal and nonfederal systems that would make use of the latest computer and communications technology.[6] The Office of Science Information Service of the National Science Foundation was to be responsible for the national center. Although the bill was not considered by any congressional committee, Congressman Pucinski's activities, which were reinforced by many recommendations for a national information system, persuaded many officials that the federal government should develop a national program for a coordinated network of sci-tech information centers. One of the active participants in the discussion was Stafford Warren, the president's advisor for the mentally handicapped. In 1964, Warren prepared a plan entitled "National Library of Science System and Network for Published Scientific Literature." Even though the plan was not seriously considered by the Committee on Scientific and Technical Information (COSATI), FCST, or by any of several congressional committees, it did add to the belief that a technologically advanced national science information system should and could be developed.

Interdepartmental Committees

Federal Advisory Committee on Scientific Information (FACSI)

Even before the amended Executive Order 10521 was issued by President Eisenhower, James R. Killian, the president's science advisor, encouraged the

The Decade of Investigation, Reorganization, and Expansion 1957-1966   57

National Science Foundation (NSF) to establish an interdepartmental committee to monitor and advise on sci-tech information activities in the government. The director of NSF, Alan T. Waterman, invited all R&D agencies to name one member each to such an interdepartmental committee. Seventeen R&D agencies selected such representatives for the Federal Advisory Committee on Scientific Information (FACSI), which was chaired by Burton W. Adkinson of NSF. The committee's recommendations were to be made to the director of NSF who would see that they were considered by the responsible governmental R&D body. This committee, which met first in early 1959, was short lived. It functioned long enough, however, to review the problems facing the federal sci-tech information agencies.[7] FACSI's only positive recommendation that was actually approved was the federal policy honoring page charges for publication in not-for-profit scientific journals. A weakness of FACSI was that its recommendations were made to the head of one federal agency. FACSI's members were reticent to suggest changes in the policies and program of one agency which were to be handled by the head of another agency, namely the director of NSF. FACSI was disbanded in 1961 at the time President Kennedy abolished all committees believed not to be essential to the operations of R&D programs.

Committee on Scientific Information (COSI)

When Jerome B. Wiesner became science advisor to the president, he gave considerable attention to the federal sci-tech information situation. As noted before, both the Crawford task force and the Weinberg panel were activated soon after he took office. In late 1961, he also established the Committee on Scientific Information (COSI) as one of the interdepartmental committees of FCST. The decision by Wiesner to have COSI report to FCST was based on the premise that the committee's recommendations would come to the attention of federal R&D officials in a position to take necessary action. To strengthen consideration of federal sci-tech information programs further, in 1962 Wiesner appointed J. Hilary Kelly as technical assistant to monitor these programs, including COSI.

COSI spent the first few months developing operating procedures that would enable it to obtain data on problems facing the federal sci-tech information agencies. In 1962 Charles Martell, deputy director of the Defense Department's Directorate of Research and Engineering, became chairman of COSI. He immediately established a vigorous committee program by selecting problems identified by COSI members as primary in importance and also insisted that agencies name persons to the committee who served as the agency focal points, which was in line with the Crawford task force's recommendation that was approved by the FCST. Martell then appointed working groups to investigate and recommend action on various problems needing interdepartmental attention. For example, the large technical information centers were hampered by the lack of standards for microfilm and microfiche. Edward Brunenkant of AEC was appointed as chairman of a group to consider this matter. The group met with interested federal agencies, representatives of manufacturers of microform equipment, and representatives of professional and technical societies. The result was the development and acceptance by the federal agencies of a set of standards that was also agreeable to both the manufacturers of and the other organizations using microform products. Another problem was the need to provide reliable estimates of funds expended by federal agencies for sci-tech information activities in response to congressional committee and executive office requests for such data. Prior attempts to collect such information had been frustrated for two reasons. First, the agencies frequently included sci-tech funds as subitems in other budget categories where they could not be readily identified, and second, some federal information agencies were reticent to compile such data lest it subject them to undesired scrutiny by the Bureau of the Budget and congressional committees. Martell, with Bureau of Budget concurrence, assigned this task to the NSF representative since for some years NSF had been preparing annual surveys of federal expenditures for R&D activities. The first try at the compilation was made by the staff of the Office of Science Information Service on data for fiscal year 1963. In 1964 the survey became a part of the regular NSF report on federal R&D expenditures. In succeeding years, the categories were redefined and broadened. For example, in 1966 expenditures for technical information activities were included for the first time.[8] In 1963 Martell instituted the practice of preparing an annual report that summarized the activities of COSI and contained brief statements on the activities of the federal agencies' sci-tech programs. This practice was continued until 1971.[9]

Other actions that were recommended and adopted by COSI are as follows: transfer of the fiscal and managerial responsibility for the Science Information Exchange to NSF; endorsement of the National Bureau of Standards proposal for a National Standard Reference Data System; and endorsement of the Commerce Department's proposal that the responsibilities of the Office of Technical Services be enlarged and that it be renamed the Clearinghouse for Federal Scientific and Technical Information (CFSTI).

A problem that COSI inherited from FACSI was the continuing pressure from members of the staff in the Executive Office of the President, some members and committees of Congress, and people and organizations outside the federal government to develop a national plan for sci-tech information activities. By the time William J. Ely succeeded Charles Martell as chairman of COSI in 1963, the pressure to develop a national plan for sci-tech information activities could not be ignored. COSI, whose name was changed to the Committee on Scientific and Technical Information (COSATI) in 1964, decided to develop such a plan. Ely appointed a COSATI Task Force on National Systems for Scientific and Technical Information. Soon after the Task Force was organized,

Ely resigned as Chairman of COSATI, and William Knox became its chairman as well as chairman of the task force. The task force soon realized a large-scale effort would be necessary to develop a meaningful program. The System Development Corporation (SDC) was given a contract by NSF that was monitored by COSI to begin developing the basis for a national plan. The first study was limited to secondary services of abstracting and indexing.[10]

Preliminary results of this study were sent to the organizations that had furnished information to SDC. The report aroused considerable criticism because the data on operational costs furnished by the federal and nonfederal abstracting services had been considered comparable whereas the federal agencies were not in position to include such costs as operation and maintenance of buildings and other similar items that were included as costs of private organizations. As a result important data on unit costs of abstracting and indexing had to be modified or eliminated. Even before this report was completed, the task force found that it needed to base its deliberation on a broader review of sci-tech information activities. On the other hand, COSATI realized that it would be unrealistic to include too many sci-tech operations. It decided, therefore, to eliminate from its consideration activities related to engineering data, formal and informal meetings, personal communications, and primary publications.

Committee for Scientific and Technical Information (COSATI)

In 1964 COSI was renamed the Committee for Scientific and Technical Information (COSATI) to reflect its concern with information services serving both research and development programs. SDC was asked to continue its study, but to limit the investigation to activities related to handling documents. The study report was issued in 1966.[11] Meanwhile, on the basis of the SDC studies, as well as the findings of previous investigations, the task force developed a program of recommendations for a national document-handling system, which was presented to FCST in 1965.[12] Although the FCST reviewed the report, there was no strong support for the program and the recommendations were never endorsed.

In addition to the National Systems Task Force other COSATI panels and working groups were investigating specific aspects of the sci-tech information activities and programs. In 1965 these included panels on operational techniques and systems, specialized information centers, information sciences technology, education and training in the sci-tech information field, and international activities in sci-tech information. Each panel developed study programs and made suggestions for the improvement of the sci-tech information activities of the federal government. In many cases, the panels identified actions that two or more agencies could adopt without further action by COSATI. In Chapter 5 later activities of this interdepartmental committee are reviewed.

Federal Libraries Committee

Since COSATI was limited to sci-tech information, it did not perform adequately as an interdepartmental coordinating mechanism for the federal libraries, most of whom had important services not related to science and technology. In 1964, a number of federal and nonfederal librarians met with COSATI to discuss its limited scope. In 1965, the Library of Congress and the Bureau of the Budget cooperated in establishing a Federal Libraries Committee with the librarian of Congress as the chairman. The Council on Library Resources made funds available for a secretariat for this committee. Although its role did not duplicate that of COSATI, there were many areas in which the two committees had similar interests. This overlap of interest is shown by the following listing of the Federal Libraries Committee's schedule of task forces: acquisition of library materials; automation of library operations; education; interlibrary loan arrangements; missions and standards; physical facilities; recruiting; and the role of libraries in information systems.

The membership of the committee consisted of representatives from the three national libraries—Library of Congress, National Library of Medicine, and National Agricultural Library—plus six representatives elected for two years each by independent agencies. The entire committee met periodically; later, an executive committee was elected that met more frequently. The task forces, however, were long-range operating arms of the committee. These groups sometimes used federal personnel to perform their assignments; in other cases outside groups or individuals were employed to carry out studies and experiments. In addition, the committee appointed groups of individuals to work on specific, relatively short-term problems such as the interpretation of Civil Service Classification guide lines, procurement, program planning, budgeting procedures, statistics, and map libraries. These groups and the committee were just beginning to operate at the end of the 1957-1966 decade.

Federal Departments and Agencies

Pressures from Congress and the Executive Office of the President affected individual departments and agencies, but their primary interest was the improvement of their R&D programs. During this period, many departments and agencies reviewed their sci-tech information operations. Although these studies often were conducted by internal working groups, outside management organizations sometimes were employed. As the result of these investigations, the R&D agencies made many organizational, policy, and program changes in their information activities.

A number of conditions caused problems for the libraries and information centers. For one, the quantity of published sci-tech literature was doubling

The Decade of Investigation, Reorganization, and Expansion 1957-1966   61

every eight to ten years. R&D findings were also being reported in an increasing number and variety of languages. For example, before World War II, English, German, and French were the three major languages of scientific communication; by 1957 Russian and Japanese rated second and third in extent of publication in many fields of science, and the predominance of English was lessening. By requiring additional translation, these changes added to the burden of information services, as did the wider distribution of R&D centers and greater variety in areas of specialization. Further, the increase in the quantity of numerical data being used added another service problem. Because of the increase in the speed of communications and travel, scientists and engineers demanded faster information services just at the time these services were being overburdened with expanding acquisitions and the need to develop services for new areas of science and technology. Finally, the effective introduction of new technologies added another dimension to the problem. This last-named factor was particularly bothersome because information specialists, systems analysis, and R&D administrators underestimated the complexity of and the time required to introduce the new technologies into operating systems.

POLICY AND ORGANIZATIONAL CHANGES

As indicated in Chapter 3, the period from 1942 to 1957 saw the development of large technical information centers in the Commerce and Defense Departments and in the Atomic Energy Commission. Other departments and agencies expanded their libraries and specialized information services and established many new ones. The succeeding decade, 1957-1966, was characterized by many organization and policy changes made in response to pressures of the accelerating growth of sci-tech information and the expanding use of this material.

The Department of Defense expanded the role of the Armed Services Technical Information Agency (ASTIA) and renamed it the Defense Documentation Center (DDC). It was removed from the R&D arm of the department and placed in the Defense Supply Agency. To strengthen its management, a well-qualified civilian, Robert Stegmaier, was appointed director. To strengthen the Defense Department information programs further, in 1962 the department established an office to be the focal point for sci-tech information in the Directorate of Research and Engineering. Walter Carlson, who was experienced in the sci-tech information field, was appointed head of this office. Other steps taken by the Department of Defense included expanding and reorganizing the departments cartographic and geographic programs by establishing large information centers in the Navy, Air Force, and Army mapping agencies. Markedly expanded information services were developed in the Bureau of Ships, the Bureau of Naval Electronics, and the Office Naval Research Laboratory, as well in the numerous R&D centers. One new service that was started with the cooperation of other federal agencies was the National Oceanographic Data Center.

The Atomic Energy Commission expanded the technical information services in its national laboratories, raised the Technical Information Section to divisional status, and broadened the latter's responsibilities to include exhibits and some educational programs. Many new specialized information centers were organized in AEC's national laboratories; a number of these were delegated to evaluate and disseminate numerical data. The Industrial Information Branch was formed as part of the AEC Technical Information Section, and TIS was also authorized to conduct exchange programs with many countries, including the Eastern bloc nations.

The Commerce Department made a number of organizational and policy changes to improve its sci-tech information services. The operations of the Office of Technical Services were moved to adequate space in Springfield, Va., where the various OTS activities were consolidated under one administration as a result of a technical survey made by Bernard Fry of the National Science Foundation.[13] Then, following the appointment of Fry as director, the role of OTS was clarified, and it was renamed the Clearinghouse for Federal Scientific and Technical Information (CFSTI). Previous to Fry's appointment, OTS had been transferred to the Institute of Technology in the Bureau of Standards, and the Department of Commerce had negotiated a contract with the Department of Defense whereby OTS organized the bibliographic information on unrestricted Defense Department reports and became the public outlet for these reports. With the endorsement of COSATI and cooperation of other major technical report centers, CFSTI began to issue a consolidated index to all federal, government-sponsored sci-tech reports and other scientific publications not published by the Government Printing Office. By this move CFSTI became the de facto sales outlet for government scientific and technical reports.

The Bureau of Standards established two new services. The first, in 1959, was a Research Information and Advisory service for Information Processing. Then in 1963, the bureau organized the National Standard Reference Data System to improve service on critically evaluated data and to strengthen U. S. participation in the corresponding international program known as CODATA.[14]

In 1965, on the Commerce Department's recommendation, Congress passed PL 89-182, which established the Office of State Technical Services. It was authorized for three years to develop cooperative programs with agencies selected by each state that would encourage the use of new technologies. The department also expanded the Weather Bureau program into the National Weather Service. This service later became part of the Environmental Sciences Service Agency, which included the Coast and Geodetic Survey and the Bureau of Commercial Fisheries. The information activities of these member bodies were

The Decade of Investigation, Reorganization, and Expansion 1957-1966   63

coordinated so that the agency was able to have a comprehensive information service on the environment as related to water and air.[15]

During this decade the Library of Congress took three steps to improve its service in science and technology. In 1958 it combined the Science Division and the Technical Information Division into a Division of Science and Technology under the direction of John Sherrod. The National Referral Center for Science and Technology was also formed with financial support from the National Science Foundation. This center was set up to provide information on organizations and individuals specializing in particular fields of scinece—that is, who was working on what. John F. Stearns became the head of this new service. The third major action by the Library of Congress related to science and technology was the establishment of a Science Policy Unit in the Legislative Reference Service (now the Congressional Research Service), with Edward Wenk directing the program. In addition to advising the Congress on matters of science policy, this unit was the Library's focal point for congressmen to receive information related to science and technology.

During the 1957-1966 period two major federal information agencies were established. The first was the Office of Science Information Service in the National Science Foundation, which was organized in 1958. This office's legislative base involved two congressional actions. When NSF was authorized by Congress in 1950, the enabling legislation included a statement that it was "... to foster the exchange of scientific information among scientists in the United States and in other countries." The NSF responded by establishing as one of its first operating units, the Office of Scientific Information (OSI), with Robert C. Tumbleson as Head. In 1952, its first year of operation, OSI made grants for support of journals in physics, translation of a Russian book on chemical thermodynamics, and publication of a proceedings volume of a symposium on Soviet science.[16] In 1955 NSF reorganized the Office of Scientific Information and Alberto Thompson replaced Tumbleson as the head.

Under Thompson, OSI initiated support programs for scientific translations and scientific publications, an information center on federal scientific information, research projects in information science, and a large exhibits program. The last-named task included organizing U. S. participation in the science exhibit for the Brussel's World Fair in 1959 and a series of exhibits on earth satellites.

When Title IX of the National Defense Education Act of 1958, which was mentioned earlier in this chapter, was passed, NSF renamed OSI the Office of Science Information Service (OSIS) under the direction of Burton W. Adkinson. The NSF expanded its support program for science information through the office and placed in it responsibility for furthering the coordination and cooperation among federal sci-tech information services that Executive Order 10521 gave to NSF.[17] Between 1957 and 1966 OSIS accomplished the following:

1. Organized programs to coordinate translations accomplished with PL-480 foreign surplus credits in other countries.
2. Markedly expanded support of scientific societies' publications;
3. Acted as a catalyst and frequently a source of funds for building up cooperative federal programs, such as the regional depositories for technical translations and reports;
4. Began a large support program of research projects and studies in information science and allied fields, including mechanical translation and computational linguistics;
5. Supported the development of new bibliographic information systems;
6. Founded the Science Information Exchange;
7. Encouraged and frequently funded increased participation of U.S. citizens in the activities of international organizations, such as the Organization for Economic Cooperation and Development, the International Federation for Documentation, the Abstracting Board and CODATA activities of the International Council of Scientific Unions, and the Pacific Science Congresses.

The OSIS was unique among the federal sci-tech information agencies in that it did not maintain an operational information service. Rather, it concentrated on assisting nonfederal sci-tech information organizations to improve their products and services, supported research and studies intended to advance information and library sciences, assisted federal information agencies to improve their programs and develop cooperative projects with nonfederal organizations, coordinated the federal sci-tech translation program supported by PL-480 surplus credits, and fostered U. S. participation in foreign and international sci-tech information organizations.

The formation of the Office of Science and Technology (OST) led to increased surveillance of federal sci-tech information programs and a need to clarify the roles of NSF and OST. Early in 1964, an exchange of letters between Leland Hayworth, director of NSF and Donald Horning, director of OST, and the president's advisor for science and technology clarified the responsibilities for coordination between OST and NSF. Following is an excerpt for Horning's letter of agreement:

> The National Science Foundation shall provide leadership in effecting cooperation and coordination among nonfederal scientific and technical information services and organizations, and in developing adequate relationships between federal and nonfederal scientific information activities. The Office of Science and Technology, with the assistance of Federal Council for Science and Technology, shall provide overall leadership of all federal scientific and technical activities, including the above.

The Decade of Investigation, Reorganization, and Expansion 1957-1966   65

A second federal sci-tech information development, initiated during this decade, concerned the newly established (1958) National Aeronautics and Space Agency (NASA). In 1960, Melvin Day was appointed by NASA to expand and improve its scientific and technical information program. After being named administrative head of NASA's Office of Scientific and Technical Information, Day reorganized and expanded the information services that NASA had taken over from the National Advisory Committee for Aeronautics.[18] He negotiated one contract with Documentation Incorporated to develop a sci-tech information processing center and another with the American Institute of Aeronautics and Astronautics for coverage of published, nonfederal literature in the space sciences. The technical operations of these two organizations were carefully coordinated so that their respective outputs could be consolidated into a variety of products. One important publication was the *STAR (Space Technology and Research) Abstract Bulletin.*[19] By 1963, the NASA technical information program was one of the largest and most effective in the federal government. NASA, in 1965, formally organized a technology utilization program based on its many dispersed activities in the space field. Other large NASA information activities included efforts in the Goddard Space Center near Beltsville, Maryland, the Man-in-Space Center in Houston, Texas, and the Jet Propulsion Laboratory in Pasadena, California.[20]

NASA's sci-tech information program can be compared to that of AEC's. Both agencies were given clearly defined missions that were, at the time, popular with Congress and the scientific and general public. Therefore, each could attract highly qualified persons who developed innovative programs. In both cases, there was a predecessor agency that each inherited. Each of these former agencies had conducted high-quality, imaginative programs. Very early in their histories, both NASA and AEC initiated technologically advanced and effective information services, and each stressed wide dissemination of information about new developments in their field.

NASA's information programs in less than six years inaugurated a computer-based service, introduced the use of microfiche as a medium of dissemination, organized a technology utilization program that involved nonfederal organizations, and initiated cooperative projects with foreign and international organizations in the space field. In addition, NASA developed projects to store, analyze, evaluate, and disseminate great quantities of data related to space exploration.

Many other departments and agencies were making policy, program, and organizational changes in their information activities during the 1957-1966 decade. For example, the Department of Agriculture set up an advisory committee to work out a plan for an integrated information system. It wished its library and other information services to be designed to be technically compatible so that information in one system could be easily transferred to another. During this period, development began on the CRIST (Current Research in Science and Technology) system for maintaining a current inventory of research and development projects funded by the department and state agencies. Also, the development of the CAIN (Cataloging and Indexing) system for bibliographic operations was initiated. In 1962 the department designated its library the "National Agricultural Library" and expanded its responsibilities to permit a more aggressive program of coordination among the state agricultural libraries and greater cooperation with foreign agricultural institutions. The Federal Aviation Agency, the Department of Housing and Urban Development, and the Department of Interior and several of its bureaus also made organizational and program changes in their information activities during this period.

OPERATIONAL CHANGES: INTRODUCTION OF NEW TECHNOLOGY

At the beginning of the 1957-1966 decade, most libraries and technical information centers had incorporated microfilm into their operations and were rapidly beginning to use microcards and microfiche. At first, all microforms were used principally for storage, but very soon these were also being used for distribution of current and back materials. The savings in storage space and distribution costs was the principle reason for this rapid acceptance of the microforms in spite of strong user objections. Accompanying the expanded use of microforms was the increased use of punchcards, step cameras, such as the Listomatic, and rapid copying machines. Once the electrostatic process had been perfected to permit almost automatic reproduction of hard copy, the rapid copier became a common method of reproducing one or more copies of technical reports, journal articles, and pages from books.

During this decade, rapid advances were also made in applying computer technology and electronic communication to sci-tech information problems. In 1958, the National Science Foundation began publication of an inventory of information systems using innovative operational techniques. A survey of two issues of the NSF-published "Non-Conventional Scientific and Technical Information Systems in Current Use" for 1962 and 1966 reveals a marked growth. In the October 1962 issue, only 87 such systems were reported, and less than a dozen were employing computers. The December 1966 issue reported 175 innovative systems of which 117 were computer based. All but 37 of these allowed some form of user access to the computer file. Most, however, relied heavily on batching inquiries for search and computer printout for the users. Effective auxiliary equipment that allowed both upper and lowercase printing and selection among several type fonts was just coming into widespread use at the end of the decade.

The introduction of mechanical and electronic devices had a major impact on the classification and indexing systems in use at the beginning of this period, which was the era of intense experimentation with systems of subject

The Decade of Investigation, Reorganization, and Expansion 1957-1966   67

characterization of informational units. The "Uniterm," the "faceted," and other new indexing and classification systems came under intense scrutiny and testing. Large projects to develop special thesauri were initiated. The Engineers Joint Council developed a thesaurus of engineering terms. The Department of Defense, with cooperation of many government agencies and private organizations, initiated "Project Lex," which was an effort to expand and upgrade the indexing terms used by DDC and other Defense Department information centers. Experiments were conducted to determine the effectiveness of different indexing systems. The new automated techniques placed a heavy requirement on more precise use of indexing terms.

Many policy, program, and operational changes that occurred during this period were made possible by the new technology. The following cases illustrate the impact of these new technologies.

The National Library of Medicine had been established in 1956, and in 1957 Director Frank B. Rogers became convinced that microfilm and rapid copying technologies were advanced to the stage where it would be more economical and efficient to fill most loan requests with copies of the requested item than to loan the original through the user's library. NLM, however, required the requester to send an order through a library and NLM sent the order to this library for delivery. To facilitate this service, the book stacks in the new NLM building were designed so microfilm cameras could be moved easily to permit filming books and periodicals. These techniques of using copies for loan and filming close to the item's location in the stacks reduced the interference with regular use of the holdings. However, as is discussed later, these service innovations subsequently caused copyright problems.

In 1958, the NLM began experiments that resulted in the MEDLARS (Medical Literature Analysis and Retrieval System) program. This project involved three major components: a computer system for bibliographic organization and analysis; a rapid composing unit that was controlled by the computer (this device became known as GRACE—Graphic Arts Composing Equipment— and was developed by Photon); and through analysis and revision of the medical literature indexing system. By 1964, MEDLARS was operating; its first product was the January issue of *Index Medicus* (although the GRACE composing unit was not used until the August issue). While the initial major output of this system was *Index Medicus* and other bibliographic publications, by the fall of 1964 the NLM was able to respond to requests to search MEDLARS' file for specific items on any biomedial topic through a service known as Demand Search Service.[21] Figures 4-1 and 4-2 show components of MEDLARS, namely, the control console of GRACE and printers and the tape drive of the Honeywell computer component.

The new computer-based system allowed NLM to play an entirely new role among the medical libraries both in the United States and abroad. Legal recognition of this change in status and program came in 1965 when Congress passed

Figure 4-1. 1964 MEDLARS Computer Components. Honeywell printers are shown left and middle foreground, tape drives in background. Courtesy of the National Library of Medicine

the "Medical Library Assistance Act" (PL 89-291) that was signed into law by President Johnson in October. This legislation enabled NLM to make its bibliographic computer service system available to other libraries. The first regional library began operation in 1965; by 1970 there were a total of eleven regional libraries. By using NLM's system, these libraries were able to give a new and more effective service to medical schools for both education and research programs.

When the MEDLARS service became available in the United States, several foreign countries requested that they be allowed to participate in the program. The NLM placed two requirements on such participants. First, countries had to agree on the location of a regional library as the NLM could not furnish the system to every country. Second, the participating institution had to send staff members to NLM for training in MEDLARS indexing techniques and philosophy and in using the system. By the end of the decade, NLM was still negotiating on

Figure 4-2. GRACE component of the 1964 MEDLARS. An operator stands at the controls of the Graphic Arts Composing Equipment (GRACE) used in the first generation of MEDLARS to compose for printing the products of NLM's MEDLARS system. This model of the phototypesetter, GRACE, operated at a speed of 3,600 words per minute, (more than five times faster than previous phototypesetters) and was driven by the magnetic tapes of the computer, which was the heart of the MEDLARS I. Courtesy of the National Library of Medicine and Armed Forces Institute of Pathology

regional locations in Europe, Asia, and South America. Further developments on this service are reviewed in Chapter 5.[22]

In less than eight years, NLM had changed from an Armed Services Medical Library attempting to fulfill a national role to the focal point of a worldwide medical library system. This brief account can hardly include all the policy, program, and operational changes that were necessary to bring the system and network into being, but does illustrate how new technological advances can be the stimulus for major change in a federal government program.

At the beginning of the 1957-1966 decade, several federal agencies were experimenting with computers and reprographic techniques. In 1950, as already noted, the Technical Information Section of the AEC began using a Listomatic camera and IBM punchcard equipment in organizing and producing its indexes. By 1960, TIS was placing its bibliographic data on magnetic tape and soon thereafter was using computer and auxiliary equipment to organize and compile its indexes and bibliographies. ASTIA had begun employing a computer for control of its inventory of current Defense Department R&D projects and in 1962 inaugurated a computer system for organization, storage, and retrieval of its bibliographic data. The Library of Congress appointed a committee of experts in 1961 to plan an automated system for the production of its bibliographic tools. This project, headed by Gilbert King, was funded by the Council on Library Resources. It was 1964 before the Library began to develop the MARC (Machine Readable Catalog) system, but the Library had begun to use computer technology for small bibliographic data files such as those used by the National Referral Center and the Science and Technology Division. NASA employed computer technology from 1961 when it contracted with Documentation Incorporated to develop its bibliographic data processing center. OTS was thrust into use of computer technology when it contracted in 1963 to prepare ASTIA's unclassified *Technical Abstract Bulletin*. Thus, by the end of this period all major sci-tech information services were involved in developing computer and reprographic programs, but most of them had done little to expand their systems into a network that would connect them with their users or with other information systems.

Little attention has been given in the preceding discussion to the complex and critically important effort to improve cataloging, classification, and indexing of the informational content of documents and data compilations. No matter how efficiently a bibliographic or numerical data system performs, it is ineffective if the identification and organization of bibliographic "tags" do not lead the user to the material of interest to him. As early as 1949, an informal interdepartmental Group for Standardization of Information Services (GSIS) began meeting to establish standards for cataloging and indexing scientific and technical reports. Representatives from the Department of Defense, Library of Congress, Atomic Energy Commission, National Advisory Committee for Aeronautics, and Office of Technical Services met frequently to try and reach

The Decade of Investigation, Reorganization, and Expansion 1957-1966

agreement on the form and order, as well as the number of bibliographic descriptors that would be needed to identify their technical reports. In addition, the group gave extended attention to the form and principles of subject indexing and classification and to a common format for the catalog card. The interdepartmental group fell considerably short of full agreement after two years of intensive work. They were able, however, to develop "bridges" between their systems so that descriptive bibliographic data could readily be translated from one system to another. One specific item of agreement was a card format, known as side-margin catalog card. Nevertheless the subject cataloging and indexing practices continued to differ enough both in principle and form to present a bothersome problem throughout the postwar period.

When the Federal Advisory Committee on Scientific Information was formed in 1959, one of the priority problems facing this interdepartmental group was selection of principle, form, and content of descriptive and subject cataloging. The FACSI subcommittee, organized to investigate and recommend on this problem, made little progress. A section of the first annual report (1963) of the interdepartmental Committee on Scientific Information (COSI) was devoted to steps being taken toward developing a unified index to scientific and technical reports. The National Science Foundation underwrote a project to develop a plan and a system for compiling and maintaining a unified index to these reports. The first experimental project was the preparation of permuted title index to government reports. This index was a repetitive listing of titles in which the significant words in each report title were placed in an alphabetical arrangement. Thus, every title that included, for example, the word "plasma" would be listed alphabetically under the letter "P" and similarly for all other principle words in all titles.

Other experiments included attempts to mechanize techniques for translating one agency's indexing terms into those of another and to identify areas of agreement and disagreement among the various indexing systems. Although these experiments were not completely successful, they did provide bases for later compilation of a consolidated index. Many other attempts were made to upgrade the indexing glossaries and make them amenable to computer manipulation. Among these were the Defense Department's efforts that produced a thesaurus of indexing terms (Project Lex); NLM's upgrading of its list of biomedical terms; and the work of the COSATI task group that produced a subject category list in 1964. This classification scheme was accepted by other federal agencies to the extent it was compatible with each one's own subject heading list.

One result of the rapid expansion of U. S. R&D was the enormous growth of the technical report literature. Some estimate of this growth is shown by the number of titles added to OTS's collections: 11,619 in 1957 and 61,100 in 1965. This fivefold increase in reports and the doubling of published sci-tech literature every ten years caused scientists and engineers to demand specialized tools because the conventional abstracting, indexing, and bibliographic products were becoming too voluminous and cumbersome to use. The information services tried to respond in a variety of ways, but the most important aids in meeting these new service demands were the new reprographic techniques that allowed for rapid duplication of copies. Also the application of punchcard and computer techniques to the organization of selective listings allowed rapid introduction of specialized publications to meet the needs of groups with highly restricted bibliographic interests. Among these products and services were the continuing current bibliographies, demand searches, technical notes calling attention to new findings, selected data listings, current awareness services, topical listings of titles, and abstracts specifically organized for mission-oriented groups.[23]

R&D PROJECTS AND STUDIES

All of the above changes in techniques, operations, and systems were made possible by the large and diversified R&D programs conducted during this decade. Some concept of their growth can be seen by a review of the NSF publication "Current Research and Development in Scientific Documentation." The 1958 volume described 53 current projects being conducted in 40 different institutions. Corresponding figures in the 1964 volume were 500 projects and in 380 organizations. A majority of these projects were funded by the federal government.

We should recall that at the beginning of this period, few reliable data were available on the characteristics and effectiveness of libraries, or on the number, size and usefulness of abstracting and indexing services. Very little work was being done on planning, design, or measurement of the use or effectiveness of systems. Publishers of primary and secondary publications assumed that these periodicals were primarily for the use of scientists and engineers in their offices and laboratories and only secondarily for consultation in libraries by students, faculty, and industrial scientists. It came as a shock to many professional societies to discover through study of their subscription lists that a majority (60 percent or higher) of their subscribers were institutions of one kind or another. Moreover, the increasing bulk of these publications began to present a storage problem for most individuals subscribers. Further the increasing number of acceptable manuscripts submitted for publication taxed the publication systems of sci-tech publishers. The growth in size in journal issues increased their costs; consequently, for financial reasons, the societies had to delay much publication just at the time the demand was increasing for more rapid dissemination of sci-tech information. Also, abstracting and indexing services began to realize that their manual systems could not cope with the increasing number of titles to be listed and indexed and at the same time meet the pressures of publication of a rapid and diversified number of products.

Numerous studies and research projects were started to investigate these problems. Many proposed solutions were put forward, including the suggestion that scientific journals and abstracting and indexing publications be replaced by other forms of communication. The secondary services established the National Federation of Science Abstracting and Indexing Services (NFSAIS) in 1958 and initiated a series of projects to gather data to aid them in improving their operations. As a result, for the first time reliable information became available on the population of journals in the various fields, on the extent of overlap coverage among the services, on costs related to various parts of the operating systems, and on similar services in other countries. In 1961, NFSAIS contracted with Robert Heller Associates for a systems and economic study. The report was released in 1963.[24] It pointed out that the major conventional secondary services in the scientific disciplines faced a combination of rapidly increasing costs and little increase in demand because the current key requirement was for specialized services organized to meet the needs of mission-oriented scientists and engineers. Since these mission information needs were constantly changing, the services should organize abstracts and indexes for repackaging in a variety of forms to meet the situation. This new organization was called "X" by the Heller Group. Federation members' systems did not allow them to conform easily to this proposed new organization, and since the new technologies had not reached the point that they could quickly or economically be adopted, the existing services saw organization X as a competitor. Although NFSAIS did submit a proposal to the National Science Foundation for a trial experiment, it was clear that the members of NFSAIS were not ready for such a marked change in their operations as moving toward an integrated network of systems.

The reprographic techniques of photo-offset and rapid and relatively low-cost copying, either from microfilm or original copy, enabled publishers, libraries, and special information centers easily to reproduce large numbers of copies of journal articles, technical reports, and sections of monographs. Use of these new techniques raised the question of the rights of authors and publishers under the copyright law. Many studies were undertaken to determine the economic and other effects of the new production techniques. Among these was the "Survey of Copyrighted Material Reproduction Practices in Scientific and Technical Fields" completed in 1962 by George Fry and Associates and issued by the National Science Foundation. This study concluded that there was widespread copying of copyrighted material, but that at that time, there was no real economic loss to the publishers and authors. The report did warn that if copying became still easier and less costly, there could be a serious impact on publishers. Similar investigations were made by commercial and not-for-profit publications. None of these studies led to a resolution of the problem, but they were useful to persons who were beginning to revise the Copyright Law of 1909.

From the beginning of this period, there was a concensus that more information was needed to devise and develop effective approaches to the informational content of publications and data collections. Among the projects pursued to meet this need were analysis of data on the effectiveness of systems and techniques of indexing and subject searching; mathematical investigations into language; experiments on techniques and systems for automatic translation; and development of plans for new mechanized information systems.

The following brief analysis of the subject categories used in NSF's "Current Research and Development in Scientific Documentation" for the years 1958 and 1964 sheds light on the characteristics and emphasis of the research, development, studies, and experimental operational projects at the beginning and near the end of this period. The 53 projects located in 40 organizations, as described in the 1958 volume, could be classified as follows: 15 were studies of methods and systems for analysis, ordered arrangements, and encoding of subject matter; 12 were R&D projects on devices whether manual, mechanical, or electronic for storing and recording information; 10 were projects oriented toward R&D of systems and devices for automatic translation of text from one language to another; and 16 were studies and analyses of information needs of scientists, of fundamental concepts applicable to documentation studies, and of contributions from other fields to solve documentation problems.

The 1964 volume organized the 500 projects being conducted in 380 organizations into somewhat different categories: 58 organizations were studying information needs and uses; 195 were investigating information storage and retrieval problems, 50 were conducting projects oriented toward achieving mechanical translation; 24 were developing equipment for information processing; 54 were working on problems of character and pattern recognition. In addition, the attention of investigators in 31 institutions was on speech analysis and synthesis; linguistic and lexiographical research related to information problems was underway in 59 institutions; and artificial intelligence and psychological projects oriented to information processing and use were being carried on in 48 and 12 institutions, respectively.

Obviously, the topical categories used in 1958 and 1964 are not comparable, since R&D had expanded greatly in the intervening six years. In fact, more projects were listed in the single 1964 category of "information needs and uses" than in all categories in 1958. Complete data are not available on the amounts and sources of funds for support of these projects in the two particular years discussed. Some indication of the growth of support of R&D is provided by the budget of the Office of Science Information Service of the National Science Foundation for these years. In 1958, OSIS obligated $425,000 for research and development projects; in 1964, $1,350,000 for research on information problems and $2,470,000 for information systems development. Other major sources of federal funds for R&D in science information included also the Office of Naval Research, the Defense Department's Advance Research Projects Agency, the Air Force Office of Scientific Research, the Central Intelligence Agency, and the Rome Air Development Center. Others that contributed to this field to a lesser degree were the Bureau of Standards, NACA, NASA, the Army Research Office, the Bureau of Education, the National Library of Medicine, the Library of Congress and the Patent Office.

CONCLUSION

The 1957-1966 decade was one of turmoil in the sci-tech information field. Never before nor since have there been so many investigations by various federal agencies and so many changes in the field. By 1957 Congress, R&D administrators, and federal executives began reorganizing the problems generated by the postwar growth of R&D—a period that can be characterized as the "Era of Information Explosion"—and began taking remedial steps. From 1957 to 1966, federal sci-tech information activities were investigated again and again by Congress, the executive branch, and by federal agencies responsible for information services—so many that this period can aptly be called the "Era of Investigations." Between 1959 and 1964, congressional committees' reviews were constantly under way, while executive branch agencies were conducting their own studies.

In spite of the time of libraries and information specialists consumed by these investigations, remarkable progress was made to enlarge the sci-tech services, to improve their quality and variety, and to meet the greater demands. Progress was made in improving cataloging, indexing, abstracting, and other bibliographical activities. Accompanying this major effort to upgrade the intellectual techniques for organizing and dissemination information were even greater efforts to develop and introduce operational innovations into the systems. As we have seen, by 1966 most libraries and information centers were routinely using such reprographic techniques as microfilm, microcard, and microfiche, and many of the larger libraries and information centers were using computers and other auxiliary equipment for sorting and organizing bibliographic and other data. Some of the more advanced centers were using computer systems in answering user requests, and a limited few were establishing networks based on computers and telecommunications.

NOTES AND REFERENCES

1. President's Science Advisory Committee, *Improving the Availability of Scientific and Technical Information in the United States*, White House Press Release, December 7, 1958, 9 pp. This became known as the *Baker Report* since William O. Baker was the chairman of the panel that prepared the report.
2. James H. Crawford et al., *Scientific and Technical Communications in the*

Government: Task Force Report to the President's Special Assistant for Science and Technology (Springfield, Va.: Clearinghouse for Scientific and Technical Information, 1962), 81 pp. This was known as the *Crawford Report.*

3. President's Science Advisory Committee, *Science, Government and Information: The Responsibilities of the Technical Community and the Government in the Transfer of Information* (Washington, D. C.: Government Printing Office, 1963), 52 pp. Alvin Weinberg was chairman of the panel that prepared this report known as the *Weinberg Report.*
4. U. S. Congress, Senate Committee on Government Operations, Subcommittee on Government Reorganizations and Internal Organizations, *Documentation, Indexing, and Retrieval of Scientific Information*, 86th Cong., 2nd sess., 1960, S. Rept. 113, 283 pp. The subcommittee was chaired by Senator Hubert Humphrey. Senate Report 15, consisting of 22 pages, was issued by the 87th Cong., 1st sess., 1961, as an addendum. In 1961 the subcommittee issued a second report: *Coordination of Information on Current Research and Development*, 87th Cong., 1st sess., 1961, S. Rept. 263, 286 pp. These became known as the *Humphrey Reports.*
5. U. S. Congress, House Select Committee on Government Research, *Documentation and Dissemination of R&D Grants: Study IV*, 88th Cong., 2d sess., 1964, H. Rept. 1932, 148 pp. The Select Committee was chaired by Representative C. Elliot.
6. See U. S. Congress, House Committee on Education and Labor, Ad Hoc Subcommittee on National Research Data Processing and Information Retrieval Center, *National Information Center*, 88th Cong., 1st sess., 1963, Hearings on H. Rept. 1946, Vol. I, Parts 1, 2, and 3, 670 pp.; Appendix to Vol. I, 508 pp. This subcommittee was chaired by Roman Pucinski. In 1964 there were additional hearings on H. Rept. 1946. Part 4 of the report was issued by the 88th Cong., 2d sess., pp. 681-893.
7. See Federal Advisory Committee on Scientific Information, *Current Status of Documentation in the United States* (Washington, D. C.: National Science Foundation, 1959), 21 pp.
8. Committee on Scientific and Technical Information, *Scientific and Technical Information Programs: Special Analysis of the President's FY 1966 Budgets* (Washington: D. C.: Committee on Scientific and Technical Information 1964), 31 pp. See page 3 for budgets for FY 1963-64 and estimates for 1965 and 1966.
9. Federal Council for Science and Technology, Committee on Scientific Information, *Status Report on Scientific and Technical Information*, Committee's annual report of June 18, 1963. See inside cover for list of members.
10. Systems Development Corporation, *A System Study of Abstracting and Indexing in the United States*, Report TM-WD-394 (PB 172 249), (Falls

Church, Va.: Systems Development Corporation, 1966).
11. Launor F. Carter et al., *National Document Handling Systems for Science and Technology* (New York: John Wiley & Sons, 1967) 344 pp. An unpublished version was given to COSATI. See also COSATI, *National Systems Task Force Summary and Recommendations* (PB 168 267).
12. Federal Council for Science and Technology, Committee on Scientific and Technical Information, *Recommendations for National Document Handling Systems in Science and Technology* (PB 168 267), (Washington, D. C.: Federal Council of Science and Technology, 1965), 19 pp.
13. Bernard M. Fry, *Survey of Technical Information Activities of the Office of Technical Services*, unpublished report to the Assistant Secretary of Commerce for Science and Technology, July 1963.
14. See David R. Lide, Jr., and Stephen A. Rossmassler, "Status Report on Critical Compilation of Physical Chemical Data," *Annual Review of Physical Chemistry* 24 (1973): 135-58, for a brief sketch of the Standard Reference Data System and the CODATA program.
15. Patrick Hughes, *A Century of Weather Service: A History of the Birth and Growth of the National Weather Service, 1870-1970* (New York: Gordon and Beach, 1970), 212 pp.
16. See National Science Foundation, Office of Science Information Service, *A Review of Science Information Activities: Fiscal Years 1951-1959* (Washington, D. C.: National Science Foundation, 1960), 26p. Pages 1-7 give a brief description of early NSF sci-tech information programs.
17. National Science Foundation, Office of Science Information Science, Annual Reports for fiscal years 1960 to 1965 were the office's reports to the director of NSF.
18. Melvin S. Day, "The Scientific and Technical Information Program of the National Aeronautics and Space Administration," paper read at the American Chemical Society meeting, Washington, D. C., September 12, 1962, 11 pp.
19. See National Aeronautics and Space Administration, Office of Scientific and Technical Information, *NASA Technical Information Bulletin No. 1*, (Washington, D. C.: National Aeronautics and Space Administration, 1962), 7 pp., for a brief statement on plans, services, and products for use by personnel in NASA centers.
20. National Aeronautics and Space Administration, *How to Use NASA's Scientific and Technical Information System* (Washington, D. C.: Government Printing Office, 1966), 25 pp., describes NASA's sci-tech information, products, and services and gives information on libraries and regional centers that give service on NASA information products.
21. Charles J. Austin, *MEDLARS: 1963-1967* (Washington, D. C.: Government Printing Office, 1968), 68 pp., is a candid exposition on the strengths and weaknesses of MEDLARS I.

78 Two Centuries of Federal Information

22. Mary E. Corning, "The U. S. Library of Medicine and International MEDLARS Cooperation," *Information Storage and Retrieval* 8 (1972): 255-64, contains a concise review of NLM's cooperative efforts with non-U. S. organizations in the use of MEDLARS.
23. See Committee on Scientific and Technical Information, *Programs of the United States Government in Scientific and Technical Communications*, Annual Reports for 1964 through 1966. Each contains brief summaries of each sci-tech information agency's activities that show the emphasis of these programs during the last three years of this period. These reports are available at the National Technical Information Service.
24. Robert Heller and Associates, Inc., *A National Plan for Science Abstracting and Indexing Services* (Washington, D. C.: National Federation of Science Abstracting and Indexing Services, 1963), 38 pp. Available from the National Technical Information Service (PB 169 559).

Chapter 5

Consolidation, Computerization, and Retrenchment, 1966-1972

As mentioned in earlier chapters, between 1942 and 1966 a many-fold increase occurred in R&D in the United States; as these programs expanded so did not only the information they generated but the efforts to upgrade techniques for organizing and disseminating the information. By 1964, however, the exponential increase of R&D had begun to level off, and by 1966 the R&D agencies were searching for ways to economize. Among the first organizations affected were the libraries and special information centers. They were called upon to economize by curtailing services and by developing cooperative programs with their counterparts.

The 1966-1972 period in the sci-tech information field is characterized by:

1. Slackening of financial and administrative support for some areas, such as nuclear science and basic R&D, but increased support for some information services in support of R&D oriented toward national societal problems;
2. Little congressional interest beyond authorization and appropriation of funds;
3. Continued increase in specialized services closely integrated with R&D programs;
4. Expanding use of microforms, computer, and telecommunication technologies;
5. Increasing interagency cooperation;
6. Retrenchment by some federal sci-tech information agencies near the end of the period. (This trend first became evident in support of R&D in information sciences and later in operations and services.)

PROGRAM AND POLICY CHANGES

During the 1966-1972 period major information centers changed their program orientation toward emphasizing service to the nonfederal community, introducing new technology, and stressing information services of interest to R&D administrators. New and cooperative programs resulted from such operational developments as on-line interactive bibliographic systems and acceptance of technical standards for microfiche and formats of material on magnetic tape.

Developments at the National Library of Medicine (NLM) exemplify the progress made in the information programs in the federal government. The NLM regional libraries and special information centers rapidly became focal points that furnished assistance far beyond the traditional reference, loan, and bibliographic services. There appear to have been two important reasons for NLM's development and expansion of services. The first was congressional interest in improving the health care for U. S. citizens; the second was the foresight and administrative skill of Martin M. Cummings, who became director in 1964. During the 1966-1972 period, the NLM became both a national and an international center for biomedical communication.[1]

The Medical Library Assistance Act of 1965 (PL 89-291) gave the NLM broad responsibility to assist the nation's medical libraries in providing effective information services for medical education, research, and practice, and as noted in Chapter 4, this legislation opened the way for NLM to become the center of a national medical library system. In 1966 NLM was authorized to establish a Toxicological Information Center, which allowed it to expand its services to physicians and others to provide quick access to information on toxic substances. The following year NLM was able to establish a R&D program in medical communications. This R&D effort was accomplished through both intramural and extramural projects and was oriented toward problems covering the total spectrum of medical communications. Under Cumming's leadership, by 1970 the NLM had developed a network of eleven regional libraries that were integrated into a system based on MEDLARS (Medical Literature Analysis and Retrieval System) and other information tools provided by NLM.[2]

In the light of the expanding role of the National Library of Medicine, the Department of Health, Education, and Welfare (DHEW) reorganized the Library in 1968 and made it a Bureau of the National Institutes of Health. Included in this reorganization was assignment to it of responsibility for the National Medical Audio-Visual Center, located in Atlanta, Georgia. Later that year Congress further expanded the NLM role by passage of PL 90-426, which established the Lister Hill National Center for Biomedical Communications. Ruth E. Davis was named first director. The Bethesda, Maryland, center's mission included the following elements: library services, specialized information services (e.g., the Toxicological Information Center), specialized medical educational services, audio and audio-visual services, and a data processing and transmission support unit.[3] Administratively, the center was organized to include a staff to work on a biomedical communications network, as well as the

NLM's Extramural Program, Specialized Services Program, and Computer and Engineering Services. Grants were made this same year to institutions in New York, Philadelphia, Chicago, and Seattle to make them units of the national system of regional medical libraries. Also in 1968, a contract was let for the development of MEDLARS II, and experiments were undertaken to obtain access to remote files in Santa Monica, California, and Cambridge, Massachusetts, and to experiment with the use of the MESH (Medical Headings Subject File) in an advanced computer system at Lincoln Laboratory of the Massachusetts Institute of Technology.

Over the years NLM had carried on an extensive exchange of publications with medical libraries throughout the world. Its catalogs, bibliographies, and indexes were widely used. As MEDLARS I became operational, NLM allowed experimental use of this system in the United Kingdom and Sweden. By 1969 the NLM had bilateral agreements with organizations in five countries—United Kingdom, Sweden, France, Germany, and Australia—and with the World Health Organization for participation in the MEDLARS program. In return, the participating countries provided NLM with biomedical documents and an index of publications for input into the MEDLARS system. These countries were responsible for the MEDLARS services in their regions. In addition, the NLM was furnishing technical support to the Regional Library of Medicine in Sao Paulo, Brazil, as an agent for the Pan-American Health Organization. By 1971, NLM was experimenting with satellite communication of its MEDLARS data to the Sao Paulo Regional Medical Library and to a network of inexpensive terminals in Alaska to furnish medical information for emergency cases. These terminals were located in villages, field-service unit hospitals, and medical centers. Other experiments underway were a microwave network among hospitals in New Hampshire and Vermont, a MEDLARS on-line service to users throughout the United States, a toxicological on-line conversational network, and an audio-visual inventory and information system compatible with MEDLARS.

Elsewhere in the National Institutes of Health and the Public Health Service, specialized information centers were established to further research and health service programs. Examples of these centers are the Air Pollution Technical Information Center in the Bureau of Disease Prevention and Environmental Control and a Drug Abuse and Addiction Information Center in the National Institute of Mental Health. The National Institute of Neurological Diseases and Blindness contracted with universities in the United States for information centers on Parkinson's disease, brain research, hearing, and vision.

The National Cancer Institute (NCI) also had several important projects underway during the 1966-1972 period. It cooperated with the Walter Reed Medical Center and Chemical Abstracts service in a development project to employ the computer to identify and retrieve information on chemical structures and substructures. One result was the development of a typewriter-like machine that could make structural diagrams for input into the computer. Other NCI projects were the establishment of a Cancer Chemotherapy Service Center, support of *Cancer Chemotherapy Abstracts,* and developments leading to an on-line computer-based information system known as CANCERLINE. In addition, NCI cooperated with the World Health Organization and cancer research organizations in other countries both in R&D projects and information activities.

During this period, information services expanded not only in the health programs of DHEW but also in its education programs.[4] Since the early part of the 1960s the Office of Education had been funding a large educational research program, and as became apparent to the researchers and administrators, some system was needed for organizing and disseminating the resulting information was essential. Among the factors that contributed to this conclusion were the problems of dispersing information among many universities and private research organizations, integrating the multidiscipline nature of the information, which ranged in subject matter from biology and biochemistry through many fields including clinical, social, and educational psychology, and—as was common to many fields of science—selecting the most important products from a large number of journals articles, reports, and monographs.

In 1965 the Office of Education established a Educational Resources Information Center program, called ERIC. The first director, Lee G. Burchinal, developed a system based on using university departments and other qualified research centers for identifying, cataloging, indexing, and servicing material in selected areas of educational research. Compatible cataloging and indexing principles and techniques allowed the products of these dispersed centers to be consolidated in a centralized data bank and distribution center. Two private organizations received contracts for the compilation of the catalogs and indexes and for reproduction and distribution of copies of the publications. In 1966 the first of many selection and information centers (ERIC Clearing Houses) was established.

The ERIC program was an immediate success. By 1968, eighteen clearing houses had been established and had selected 8,800 documents to add to the ERIC collection. An abstract and index bulletin titled *Research in Education* was started in 1965 and by 1968 had 4,550 subscribers. In 1968 the ERIC reproduction and distribution center sold 2.4 millions microfiche cards and 9,000 copies of cited publications. In addition, ERIC prepared and distributed bibliographies and interpretative summaries in selected educational research areas and began a printed *Current Index to Journals in Education.* In 1971 ERIC began a service for making computer tapes of its resources available and negotiated a contract for developing an on-line interactive query service of its central files. A 1971 study on the use of ERIC by Bernard Fry indicated that 191,000 persons were using *Research in Education* each week and 91,300 persons consulted *Current Index to Journals in Education.*

During the 1966-1972 period, other new programs were generated by the

Consolidation, Computerization, and Retrenchment, 1966-1972   83

need for research on national problems and were funded by NSF's Research Addressed to National Needs (RANN). One of these was the AEC's Oak Ridge National Laboratory program to develop an Environmental Information Center. Five major data bases were organized on the following topics: toxic materials in the environment, social sciences, regional modeling, energy, and material resources and recycling. This facility is an example of a specialized information center that is both an integral part of a research program and a source of information for other organizations and investigators working on similar problems.

Another example of program changes and rapid development of information activities are those of the National Aeronautics and Space Agency (NASA). As noted in Chapter 4, NASA developed the manual bibliographic information system it inherited from the National Advisory Committee on Aeronautics (NACA) into one based on computer and photographic technologies, which it continued to expand and refine during the 1966-1972 period. NASA also acquired large numerical data bases from NACA. To this large and valuable collection of engineering and other data, NASA rapidly added additional numerical data generated in its research laboratories and from satellite and space probes. To make these data available to scientists and engineers in its own laboratories and in university and other laboratories doing work for NASA on contract, NASA devised advanced techniques for storage, organization, and dissemination of these data. Another NASA information activity that was expanded during this period warrants comment here. Prior to 1966, NASA had developed an exchange of publications and bibliographic data with its counterparts in Europe, one of which was the European Space Research Organization (ESRO). During the 1966-1972 period, NASA and ESRO began a cooperative program based on the NASA/RECON system (RECON is formed from the first letters of the two words that describe its distinguishing feature—REmote CONsole). NASA agreed that ESRO could employ the NASA contractor, Lockheed Corporation, to develop the latter's version of RECON so that data elements, although not identical, could be transferred from one to the other through a conversion process. Thus, NASA developed an effective cooperative program with the European community. Within the United States, NASA and the U. S. Geological Survey cooperatively developed a satellite earth resources mapping and analysis program. This activity generated quantities of data, photographs, and other forms of information that required implementing a new program in the Geological Survey to make the material available to investigators in U. S. research organizations, federal and state agencies, and foreign countries. The new types of material also required the Geological Survey to develop special techniques for storage.

Near the end of the 1966-1972 period, in 1970, the National Oceanographic and Atmospheric Agency (NOAA) was formed by the consolidation of programs from several agencies. NOAA reorganized and expanded the programs for collection, organization, and dissemination of environmental information. These were placed under an administrative unit called the Environmental Data Service.

It includes five centers: The Environmental Sciences Information Center, the National Climatic Data Center, the National Geophysical Data Center, the Aeronomy and Space Data Center, and the National Oceanographic Data Center. These centers had actually been established much earlier, although not under these specific names. For example, compilation of climatic data had been part of the Weather Bureau's program since its beginning in the early part of the nineteenth century; the geophysical data bank had been built up for more than a century by the Coast and Geodetic Survey, and oceanographic data had been collected by both the Coast and Geodetic Survey and the Navy. In the 1960s, these data had been consolidated into the National Oceanographic Data Center.

This trend toward consolidation and coordination of sci-tech data and information activities represents a policy change that was evident in other federal agencies and departments, such as NASA, the Geological Survey, the National Bureau of Standards, the Census Bureau, and the Departments of Defense, Transportation and Housing and Urban Development. Each of the last two departments had inherited a number of libraries and information activities from constituent agencies, and in 1971 the Department of Transportation established an Information Policy and Program Council to review and recommend action leading to more effective information services within the department, while HUD had begun to consolidate its numerous library units into a coordinated system in 1970.

Another policy change that began in the 1966-1972 period was wider application of charges for government information products and services. In 1969, the Department of Defense required its information analysis centers to initiate a schedule of charges for their services. The department consolidated the fiscal and policy management of these centers in the Defense Supply Agency, where the Defense Documentation Center (DDC) was located. AEC and NASA began charging for microfiche and some other services. This policy change was stimulated in large part by the Office of Management and Budget, which was influenced by for-profit organizations insisting that federal information services assess a reasonable charge for their output. The federal agencies were receptive to this policy because their budgets were either growing at a slower rate than before or were actually smaller. In most cases this policy was applied to agency contractors and grantees as well as to private individuals and organizations.

In line with this new policy, in 1971 the Department of Commerce strengthened and expanded its Clearing House for Federal Scientific and Technical Information (CFSTI) to become the National Technical Information Service (NTIS). William Knox was appointed director. The NTIS became a national distribution point for the government's sci-tech information products in many varied forms, instead of being merely a sales outlet for unclassified federal sci-tech reports and allied publications. The NTIS also developed a cooperative program with the Office of the Superintendent of Documents (the sales arm of the Government Printing Office) to coordinate the activities of the two agencies. Prior to the formation of NTIS, the CFSTI had been transferred from the National Bureau of Standards to the Office of the Assistant Secretary of Commerce for Science and Technology as a move to strengthen its administrative support.

Another significant change occurred in the federal sci-tech information structure in 1971. The Committee for Scientific and Technical Information (COSATI) was transferred from the Office of Science and Technology in the Executive Office of the President to the National Science Foundation. Prior to this administrative move, COSATI panels and task groups prepared numerous reports and recommendations, and COSATI had the following special studies prepared. In 1967, a contract was let to the American Institute for Research for a study on *Exploration of Oral/Informal Technical Communications Behavior* (PB 669 586), and in 1968 COSATI commissioned an investigation of data activities by Science Communications, Inc., entitled *Study of Scientific and Technical Data Activities in the United States.* COSATI co-sponsored and its chairman participated in the establishment of the National Academy of Sciences/National Research Council's Committee on Scientific and Technical Communications. This committee heard testimony of many experts and made a number of studies on different aspects of the sci-tech communications problem. In 1969 its report, *Scientific and Technical Communications—A Pressing National Problem and Recommendations for Its Solution*, was issued as NAS Publication 1707 (see Chapter 7 for further discussion). The new move restricted COSATI in several ways. First, the chairman was no longer a staff member in the Executive Office of the President and therefore could not be as effective as he had been in presenting the executive agencies sci-tech programs to the Office of Management and Budget, as well as to other units in the Executive Office of the President. Second the chairman was now responsible for a particular agency's sci-tech information program and could no longer act as an unbiased representative of the federal sci-tech information community.

While no major congressional investigations were carried out during the 1966-1972 era, in 1972 a study group under the chairmanship of Martin Greenberger of the Johns Hopkins University was charged to review and recommend on the federal government's sci-tech information programs, which created uncertainty as to the direction of the programs. Throughout the period, many of the agencies made studies related to their own sci-tech information programs that resulted in the program changes described in this section and the technical advances outlined in the following one.

OPERATIONAL AND TECHNICAL ADVANCES

Federal agencies have often been criticized for their slowness in adapting new technologies to their operational systems and even more frequently have been faulted for not introducing coordinated and integrated information systems. However, at the beginning of the 1966-1972 period, most federal library and special information systems were based on very large bibliographic and data files that presented unique technical and organizational challenges, and many of the agencies had catalogs, indexes, and data banks containing hundreds of thousands of entries. These would have required computer memories and storage banks with capacities of billions of bits which was not then economically feasible. In addition, the computer software and auxiliary equipment generally available at that time did not allow for economical and smooth transfer of these manual files and operations into partially or fully automated systems. Translation of one of these files into a computerized bank and the development of the system to use the bank represented a tremendous outlay of man-hours and funds. More importantly there was no agreement on just what would constitute an effective and efficient system for organizing and servicing these large data collections. Therefore, most information agencies first adapted new automated systems to handling their own current inputs and then selectively added sections of older files. Insuring that the numerous user centers could handle the information in the format of the new system entailed negotiations and training on the part of the federal information center. The NLM and Library of Congress systems development programs are illustrative of this step-by-step approach. Each agency developed training programs to acquaint user personnel with the technical and intellectual characteristics of the computerized systems.

During the 1957-1966 period, these problems were partially overcome. Communication, computer, and reprographic technologies were advancing at a rapid pace. The second generation of computers was in general use at the beginning of this period and the third soon became readily available. Introduction of larger computer memories, more versatile and faster terminals, more effective input devices, and almost a revolution in languages to control computer routines made this an era of rapid change. Even more significantly, as a large body of trained personnel who knew and understood the potential and the limitation of the new technical devices for bibliographic and numerical data handling developed, many information services moved rapidly from experiments to operations. To illustrate the advances made in the 1966-1972 period, a few examples are given below rather than complete chronological presentation of the introduction of the new technology.

As reported earlier, NASA had inherited NACA's high-quality, manually operated, sci-tech information system, which proved to be entirely inadequate for the rapidly expanding space research and development program, and NASA immediately adopted computerized techniques for organizing and searching its bibliographic files. It reorganized the cataloging, indexing, and dissemination operations so that by 1963 NASA had a tightly controlled operational system, even though some elements were operated by private organizations while others

Consolidation, Computerization, and Retrenchment, 1966-1972   87

were geographically dispersed within NASA. NASA carefully developed a thesaurus of indexing terms that reflected current usage of scientists and engineers. NASA also worked closely with DDC, AEC, and CFSTI to develop common standards or at least compatible or easily translatable techniques for cataloging and indexing.

Microfilm, microfiche, and rapid reproduction techniques were part of the NASA system. By 1966 NASA sci-tech operations were ready for the development of a large-scale on-line system, and a study looking to that end was initiated. By 1968, NASA had completed this study and design and had prepared the necessary software for its RECON system. At first it was used for internal operations. In 1969, seven stations were on-line with the ability to interrogate a file of a half million citations. The NASA team had worked with DDC and AEC so that NASA had access to their files and was cooperating with AEC to permit it to introduce the RECON system into its operations. NASA also worked with its European counterparts to make certain that their operation information systems would be compatible with its own, and more importantly that changes in one system did not create operational problems with other foreign or domestic user systems.

A successful information service or technique of one agency was often adopted by other agencies as exemplified by AEC's use of NASA's RECON system. Another information service involving these two agencies was their joint program of "Tech Briefs." NASA early initiated activities designed to transfer the new technology it developed to American industry and established a Technology Utilization Program through which the usefulness of its large R&D effort could be demonstrated to American society. The agency experimented with many different techniques to effect this transfer, including establishment of technology utilization centers as well as presenting demonstrations of possible uses at meetings. One innovation was the preparation of brief summaries of new technical findings. These "Tech Briefs," as they were called, described possible applications. AEC was also interested in the transfer of its new technology to American industry, and beginning in 1967 the two agencies cooperated in the preparation and distribution of "Tech Briefs" and thus provided a valuable information service for smaller industrial establishments that could not afford their own R&D programs.

The information systems of many federal sci-tech agencies followed a development pattern similar to that of NASA. As these operational systems evolved, the agencies identified areas in which they could cooperate. Thus AEC, NASA, and DDC exchanged bibliographic data—first in card form, then as computer printouts, and later as computer tapes. The Office of Technical Services was anxious to receive other agencies' documents and associated cataloging and indexing information, but not until CFSTI (successor to OTS) had perfected its mechanized operational system, through cooperation with DDC, was it able to make realistic use of the products of other agencies. Only when the data could be manipulated in the computer did integration of bibliographic data from several systems become operationally useful.

REDUCTION OF R&D SUPPORT FOR LIBRARY AND INFORMATION SCIENCES

Federal government support of R&D in the information field continued to expand during the first half of the 1966-1972 period. In the latter half, funding support began to level off and in some cases to decrease or disappear. There were two groups of funding sources. One group, composed of thirty-seven agencies, had responsibility for large libraries and data or special information centers that served their organizations. Their support of R&D was mission oriented and tailored to their particular needs. In many instances, however, their R&D findings and their operational experience with the new techniques were useful to other organizations. The second group specifically had missions to advance the communications and informations sciences for the benefit of the government and the nation. Among the major federal agencies in the latter group were the National Science Foundation, the Air Force Office of Scientific Research, the Rome Air Development Center, the Office of Naval Research, the Army Research Office, and the Advanced Research Projects Agency. From these six agencies came over 60 percent of the support for the federally funded projects listed in *Current Research and Development in Scientific Documentation* (*CRDSD,* no. 15) published by NSF in 1969.[5] This volume listed projects underway during the first six months of 1968. Three other agencies that gave support to a broad spectrum of R&D projects even though their interests were limited to specific fields were the National Institutes of Health and the National Library of Medicine, which had responsibility to advance the state of biomedical communications and to improve the nation's medical library services, and the Office of Education, which under its general mission for improving U. S. education had a support program to upgrade libraries and other information, related to educational research.

The change in emphasis in R&D in information and library sciences between 1966 and 1968 is shown in Table 5-1. The data for 1966 and 1968 show a 22 percent increase, from 640 to 783, in the total number of R&D projects listed in *CRDSD.* Growth in federal R&D funds for information sciences was the same percentage. Comparable data on R&D projects for the 1969-1972 interval is not available. The data on individual projects listed in Table 5-1 show significant changes in the emphasis for the R&D activities. For example, there was a slackening of projects on "information needs and uses," part of which came from a change in interest of investigators. More importantly, however, federal agencies' priorities changed. Many federal agencies had reached the stage in their

Table 5-1 Project Support by Selected R&D Categories

Category	Number of Projects	
	1966	1968
Information Needs and Uses	61	43
Information Storage and Retrieval	282	489
Machine Translation	36	24
Linguistic and Lexicographic Research	109	129
Equipment Development	22	0
All Other	130	100
Total	640	783

Source: Data from Harold Wooster, "Current Research and Development in Scientific Documentation," *Encyclopedia of Library and Information Science,* vol. 6 (New York: Marcel Dekker, Inc., 1971) Table 2, p. 340.

information systems development where they no longer felt the need for data on "information needs and uses," but instead for data on "information storage and retrieval," as evidenced by the 73 percent growth in the number of such projects from 289 to 489. The increasing use of computers and mechanical and photographic devices for storage, organization, and search of information required a better understanding of langugage and improved indexes—that is, thesauri and key word lists—for use with mechanized fields. This interest is shown by the growth of projects in the category of "linguistics and lexicographical research," which exhibited itself prior to 1966 and continued to rise from 109 projects in 1966 to 129 in 1968. Another factor, that led to the rise, was new research priorities by some information scientists. Many researchers, who initially were interested in trying to achieve machine translation, reoriented their research to investigation of language structure, problems of machine abstracting and indexing, and other techniques to improve searches with computers or mechanical devices.

Fiscal data on information sciences R&D for the 1969-1972 interval is not organized topically within the R&D categories, but a brief review of what is available shows changes. For example, the 22 percent increase in R&D funds noted above for the 1966-1968 interval contrasts with a 9.4 percent increase for the 1969-1972 interval. The increase in funds for the 1966-1972 period was $21.7 million, or 45 percent, which can be contrasted with the 1960-1965 period when the increase was $45.1 million, or 155 percent. During the 1966-1972 period the inflation rate was higher than for the previous periods, so in real terms, growth slowed even more than the figures suggest. Data for R&D in information sciences were quite incomplete until 1964, but the 1964-1966 data show an increase was $35.5 millions, or almost 156 percent. Thus, support for R&D in information sciences rose very rapidly just prior to the 1966-1972 period, continued through 1968, and then leveled off markedly.

Table 5-2 shows the changes in obligations for federal sci-tech information

Table 5-2 Federal Obligations for Sci-Tech Information Activities, 1966 and 1972 (in millions of dollars)

Category	1972	1966	% change
Total	$419.4	$277.7	+51
Publication and Distribution	116.7	82.7	+41
Documentation, Reference, and Information Services	196.5	124.6	+58
Symposia and Audio-visual	35.5	22.4	+58
Research and Development	69.7	48.0	+45

Source: National Science Foundation, *Federal Funds for Research, Development and Other Scientific Activities,* Vol. 16 and Vol. 22 (Washington, D. C.: Government Printing Office, 1967, 1974), 248 p. and 61 p. Data for 1966 was taken from table in Vol. 16, p. 223, and data for 1972 was taken from tables in Vol. 22, p. 38.

activities during the 1966-1972 period. "Publication and distribution" funding did not rise as sharply as the next two categories listed in the table because many agencies curtailed their free distribution or made distribution in microfiche form, which materially reduced the cost per item. In addition, more agencies began furnishing NTIS with copies of their material for reproduction and sales to requestors.

During the period 1966-1972, military agencies were required to limit their support of all R&D to defense-related projects. Thus the Office of Naval Research, the Air Force Office of Scientific Research, and the Advanced Research Projects Agency reduced the scope and size of their support of R&D in information sciences. As was related earlier, NSF's support remained at about the same level. However, during this same period the Library of Congress, the National Agricultural Library, and the National Bureau of Standards were expanding their R&D efforts. New agencies were beginning to fund projects in this area. Among these were the National Oceanographic and Atmospheric Agency and the Environmental Protection Agency.

CONCLUSION

The 1966-1972 span is characterized by program and policy changes and contrasting developments in growth. Most large federal information agencies inaugurated advance computer and telecommunications bibliographic systems, even though growth of fiscal support decreased. New or enlarged information services were organized such as ESSA's Environmental Research Information Services, the change of the Clearinghouse for Federal Scientific and Technical Information into the National Technical Information Services with its enlarged program, and the beginning of the Environmental Protection Agency's information activities. A number of cooperative projects between information centers were started. For example, the three national libraries began a joint project to develop better control over serial publications; NASA and AEC cooperated in the use of the former's RECON system; most technical report-producing agencies increasingly relied on FCSTI and later NTIS for sale of their information products; the Standard Data Reference System under the National Bureau of Standards leadership began to operate; and there was increasing exchange of information in magnetic tape and microfiche form among information centers.

NOTES AND REFERENCES

1. National Library of Medicine, *Historical Chronology and Selected Bibliography Related to the National Library of Medicine* (Washington, D. C.: National Library of Medicine, History of Medicine Division, 1971), 20 pp. See pages 13 and 14 for events during the period covered by Chapter 5.
2. National Library of Medicine, *The National Library of Medicine; Progress and Services: Fiscal Year 1973,* DHEW Pub. No. (NIH) 74-286 (Washington, D. C.: U. S. Department of Health, Education and Welfare, Public Health Service, National Institutes of Health, 1974), 40 pp. This report gives a concise summary of NLM's products and services; it also shows the organization of NLM and members of the board of regents.
3. U. S. Department of Health, Education and Welfare, Public Health Service, National Institutes of Health, National Library of Medicine, *The Lister Hill National Center for Biomedical Communications: Report to Congress, May 1974,* DHEW Pub. No. (NIH) 74-706 (Bethesda, Md.: National Library of Medicine, 1974), 18 pp. This report gives a brief history of the development and the present program of this center.
4. See Committee on Scientific and Technical Information, Annual Reports for fiscal years 1966-1972. These contain brief summaries of each federal agency's sci-tech information activities. These prove useful in getting a perspective on the federal sci-tech information activities for these years. Most federal annual reports of departments and agencies give little attention to their sci-tech information programs. The annual reports are entitled *Progress of the United States Government in Scientific and Technical Communication* and are available from the National Technical Information Service.
5. National Science Foundation, *Current Research and Development in Scientific Documentation,* Vol. 15 (Washington, D. C.: Government Printing Office, 1969), 741 pp.

Chapter 6

Federal R&D Programs and Studies

This chapter concentrates on the federal government's role in R&D in library and information sciences and in studies that were important in determining governmental courses of action. The reader must recognize, of course, that most of the R&D and many of the studies were performed outside the government under grants or contracts.

In previous chapters, R&D in information sciences and government-sponsored program and policy studies were considered in the context of each chapter's time period. This chapter presents an overview of the government's efforts in R&D and studies from 1942 through 1971. The reader should be aware that R&D cannot be compartmentalized into time periods or scientific or engineering disciplines. As an illustration, R&D on mechanical and photographic devices and systems for organizing and distributing sci-tech information had been initiated before 1942 and efforts to refine these processes continued beyond 1972. In fact, the Department of Agriculture as early as 1911 was experimenting with use of microfilm copies in lieu of loan of the original document, and in 1972 efforts were still being made to improve microfilm readers and the quality of film.

During the 1942-1972 interval, the information community experienced marked changes in organizational structure, available technology, amount and sources of funding, and administrative and program alignments. From 1942 to the early 1950s, federal information services were faced with burgeoning federal scientific R&D that caused a rapid growth of sci-tech literature. This growth, part of which was a result of the release of U. S. publications that had been unavailable during World War II as well as documents captured from the Germans and Japanese, created a heavy burden on sci-tech information services. Federal agencies as well as private organizations began to explore ways to handle the increasing volume as well as to give service to a rapidly growing and increasingly varied clientele.

Federal agencies initiated or accelerated their efforts to make more effective use of available photographic and mechanical technologies, and this era saw expansion and improvement of photoduplication services, widespread and new uses of punchcard techniques, and a frantic search for technology that would replace the "wet-process" of facsimilie reproduction. Federal agencies, either within their own organizations or by contracts, stimulated the development and operational use of punchcard systems for organizing indexes, abstracts, and classifications, and the development of the electrostatic process for facsimile reproduction that is exemplified by the Xerox "Copyflow" machine. Offset printing and multilith reproduction became more refined and widespread in use.

The formation of many new sci-tech information centers both within and outside the government during this period stimulated the need for better coordination and standardization. One effort to achieve coordination and standardization was an informal group called Group for Standardization of Information Services (GSIS). This group met many times during the late 1940s and early 1950s to explore ways to coordinate, improve, and standardize techniques of cataloging, indexing, and publication. The only success was agreement on the format and catalog elements of entries on catalog cards or in announcement bulletins. Other problems investigated in this period were the use of microcards, methods of indexing and classification, and uses of these techniques in photographic and punchcard systems.

As R&D increased, scientists and engineers began to experience more and more difficulty in acquiring needed information. R&D administrators were made aware of this problem, and by 1955 several of the federal R&D agencies had assigned a person to stimulate and administrate R&D in the information sciences. The expectation was that the resulting new technology in information processing would resolve some of the sci-tech information problems.

The federal information sciences R&D administrators felt the need to be informed on current R&D information sciences. In 1955 they began to meet informally to discuss the interests and activities of researchers in the field and to coordinate their support programs. These meetings, although informal in nature, became regularly scheduled events and the group, which became known as the Interagency Group for Research on Information Systems (IGRIS), exerted a strong influence on the character and direction of federal support of R&D in information sciences.[1] A description of IGRIS's method of operation is helpful in understanding the pervasiveness of its influence. Representatives of federal agencies who were not regular participants were invited from time to time to discuss their agency's information R&D effort. Members of IGRIS developed effective liaisons with people and organizations in foreign countries, and the foreign groups were invited to present their R&D in information sciences to the IGRIS meeting. IGRIS closely monitored information sciences meetings both in the United States and abroad. Finally, IGRIS invited U. S. investigators to present their findings to the group.

The need to be informed of current R&D projects in information sciences stimulated IGRIS to encourage the National Science Foundation to initiate a serial publication containing descriptions of projects and citations pertinent to each project's area of investigation. *Current Research and Development in Scientific Documentation (CRDSD)*[2], as this serial was called, became a major reference tool between 1956 and 1968 for information sciences investigators and administrators.

Turning again to discussion of sources of funding and the character and trends in information sciences R&D during the 1956-1969 period, the author found Harold Wooster's in-depth analysis of *CRDSD*[3] to be good description of this period. Wooster, by count of numbers of R&D projects listed in each volume of *CRDSD*, identified the following agencies as the major sources of funds: National Science Foundation, Rome Air Development Center of the Air Research and Development Command, Office of Naval Research, Air Force Office of Scientific Research, U. S. Army, National Institutes of Health, Office of Education, and several agencies in the Department of the Air Force other than the Office of Scientific Research. Several agencies in addition to those mentioned by Wooster were supporting information science and technology research. Among these were the National Library of Medicine, the Patent Office, the National Bureau of Standards, the Library of Congress, the National Aeronautics and Space Administration, and the intelligence agencies.

The change in interests of investigators and the growth of the field can be shown by the total numbers of projects and the topics under which these were listed in three selected volumes of *CRDSD*. Volume 1 (1957) described 44 projects under the following headings: organization of information, equipment for storage, and retrieval; special studies; and mechanical translation. Volume 7 (1960) included 159 projects that were organized under the four headings of Volume 1 as well as these additional categories; information requirements and uses; research on information storage and retrieval; character and pattern recognition; and other research. By the time Volume 15 (1969) was compiled, there were 785 projects organized under twelve headings: information centers, information programs, theoretical studies of national plans and networks, and requirement studies; information use and communication patterns; language analysis; machine translation; indexing and classification theory and practice and thesaurus development; automatic content analysis; retrieval systems; publication and announcement systems; library operations; performance, analysis, and evaluation; pattern and speech recognition; and adaptive and interactive systems.

CRDSD fell victim to the budget restraints that the federal information activities experienced between 1968 and 1972; therefore, there are no comparable data for that period. R&D in information sciences during this latter period continued along the line indicated in Volume 15 of *CRDSD* but at a reduced level. One noticeable trend during this 1968-1972 period was the emphasis that federal information agencies gave to improving their own systems. One exception to this trend was that of the National Science Foundation, which expanded

96   Two Centuries of Federal Information

its support to scientific and professional societies for the planning and development of advance information systems.

The following sections of this chapter describe information R&D efforts of one federal agency, the Patent Office, and analyzes the impact of federally sponsored program and policy studies. The Patent Office's R&D program was selected because it was one of the first agencies that tried to incorporate new technologies into its information activities, its program was markedly affected by congressional and departmental directives, and it demonstrates an agency's attempts to develop inter-agency cooperation in R&D as well as interactions with its foreign counterparts.

PATENT OFFICE[4]

An early federal information sciences R&D effort was that of the U. S. Patent Office to use mechanical and electronic searching systems in the examination of patent applications. Its early initiation, and the involvement of many individuals and organizations both within and outside the government, make this Department of Commerce project an important element in the federal government's efforts to develop programs and policies related to sci-tech information activities.

Between 1948 and 1951 a small staff in the Patent Office prepared a co-ordinate index of patents related to the "composition of matter" and developed a system for search of this index. This project showed potential for searching patents with greater speed and efficiency than the manual system then current. However, development and operating costs were so very high that the Patent Office discontinued the program. This experimental system aroused interest among federal officials in both the legislative and executive branches, however, and many persons were beginning to predict that sci-tech information systems based on the newly available electronic computers would soon replace many of the manually operated ones.

In 1954, when the Senate Appropriations Committee acted on the Patent Office's 1955 budget request, it directed that "an aggressive and thorough investigation be made as to the possibility of mechanizing search operations and installation of up-to-date types of equipment in order to modernize, insofar as practicable, the Patent Office's operations." Congress recognized that the Patent Office maintained and used some of the largest and most complex files in the federal government.

In response to the Senate Appropriations Committee's directive, Secretary of Commerce Weeks appointed a committee to investigate and recommend on modernizing the Patent Office's information operation. This committee, which was representative of a broad spectrum of interests, was chaired by Vannever Bush, formerly director of the Office of Scientific Research and Development and originator of the idea of "Memex." The committee's report in 1955 recommended that:

> A. The Patent Office should put machine searching of "composition of matter" on an operation basis. [This was the system developed by the 1948-1951 project staff.]
> B. A Research and Development Unit should be established in the Patent Office, and
> C. The National Bureau of Standards and the Patent Office should undertake a joint program to stimulate and develop machines and techniques specifically adapted to the Patent Office.

The reader should recall the "state-of-the-art" at the time this effort was initiated: existing computers had small core memories, operating speeds were slow compared to those of today, adequate auxiliary equipment was just becoming available, and programming languages were in the developmental stage.

The major Patent Office requirement was for an effective and rapid search system capable of handling many types of sci-tech information, including that in graphic form. Thus its efforts concentrated on developing searching techniques and systems. The early efforts were handicapped, however, by the necessity to limit severely the searching vocabulary and to restrict the complexity of search strategies because of the limitations of technology. The first projects were directed toward developing small systems oriented toward discrete segments of its files, and even these search strategies were limited to searching a portion of the information in the file. For example, one early system that was developed to search patents on steroid compounds was able to identify the compounds but could not select the processes involved in their creation. Despite this limitation, however, this search system proved useful to the patent examiners and was subsequently refined and expanded to provide a more comprehensive search system for this class of patents.

In 1956, the first response to the Bush Committee's recommendations was a cooperative systems development program, known as HAYSTAQ, between the Patent Office and the National Bureau of Standards (NBS). (The author could find no meaning for the acronym.) This effort was directed toward the development of a mechanized system using the NBS experimental computer, SEAC (see Figure 6-1). While the proposed system would be able to search all the contents of a technical document in any field, patents related to chemistry were selected for the first experiment since participants in the project had been trained in chemistry. The tasks involved for the Patent Office in this systems development program were the identification of questions used by patent examiners to obtain information and the design of a large file for use in the SEAC computer to index, organize, update, and search. At the same time, NBS was experimenting

Figure 6-1. Console and main frame of the National Bureau of Standards' Experimental computer, SEAC, which was used in an early NBS and Patent Office joint systems development program called HAYSTAQ

with its computer to design components and strategies to meet the requirements for storage and manipulation of the large file.

Other projects associated with HAYSTAQ until 1959 included: coordinate indexing of patents dealing with earth-boring tools; an experiment by Victor H. Yngve of M.I.T. in the application of machine translation techniques to develop suitable language structure for possible use in mechanized retrieval; an experiment in digital storage of graphic information; and the improvement of the Rapid Selector for use in a system to select documents on microfilm pertinent to a search query. By 1959, the Commerce Department and Congress had become impatient with the progress of the Patent Office and the NBS toward developing a mechanized search system. The secretary of commerce requested the National Academy of Sciences to examine the operations of the Patent Office. Mervin J. Kelly, president of the Bell Telephone Laboratories, was named chairman of a committee that reported to the secretary of commerce in 1960. The report indicated that much effective work had been accomplished by the R&D program and made the following recommendations:

1. A comprehensive systems engineering study of the Patent Office be made by competent outside organization;
2. The R&D program shift its direction from a shot-gun approach to a more concentrated attack in the area of composition of matter;
3. The R&D staff be augmented by specialists in the fields of physics, mathematics, and linguistics;
4. The large, long-term projects requiring a broad spectrum of talents be placed at the Bureau of Standards where the climate was more conducive to attracting necessary specialists;
5. A committee advisory to the Commission of Patents and the Director of the National Bureau of Standards be appointed to monitor and advise on the R&D program.

The commissioner of patents appointed a panel in 1960 to suggest how the Kelly recommendations should be implemented. This panel, which was chaired by Gilbert W. King, recommended in 1961 that the NBS should seek funds for a substantial R&D program in information storage and retrieval. It was urged also that the Bureau establish a Clearinghouse and Coordinating Center for information retrieval, with the Patent Office continuing to develop small retrieval systems suitable for searching sections of the patent files. The NBS obtained funds for only a limited R&D program. The National Science Foundation funded the initial clearinghouse activity at the Bureau with the understanding that the NBS would take over its support when it became operational. The Patent Office was not successful in obtaining increased funds from Congress for its development program.

When a new commissioner of patents took office in 1961, he asked Director A. V. Astin of NBS and J. C. Green, director of OTS, to advise him on steps needed to implement the Kelly and King groups' recommendations. The commissioner also consulted with the staff of OSIS of the National Science

Foundation regarding actions he should take to improve the Patent Office's R&D program. One result of these consultations was the appointment of R. A. Spencer as director of R&D in November, and Ezra Glazer as consultant with oversight of all R&D efforts related to the Patent Office program. In May 1962, Ezra Glazer was appointed assistant commissioner for R&D. During the next ten years the Patent Office remained unsuccessful in its quest for increased R&D funds. In spite of this limitation, its R&D program conducted useful projects for improving searching techniques, and the HAYSTAQ project developed enlarged and better searching strategies and also carried out various statistical and psychological studies.

In 1961, at the time of the celebration of the 125th anniversary of the Patent Office, an informal organization of representatives of patent officials working on information retrieval problems was started. This Committee for International Cooperation in Information Retrieval among Examining Patent Offices (ICIREPAT) met infrequently but did develop some useful cooperative projects. In 1962 the Patent Office, with NSF funding, initiated a program of visits of guest experts from other patent offices. This program, which was under the sponsorship of ICIREPAT, included sharing techniques and systems of indexing. Plans were underway in 1964 to develop guidelines for searching strategies and methods to test the effectiveness of different searching systems. The goal was achievement of common searching systems for international use. U. S. participation was severly limited in 1965 when Commission E. Brenner reduced the R&D program, and Ezra Glazer resigned from the Patent Office. In 1968 ICIREPAT became a committee of the Paris Union.

In sum, the Patent Office's program between 1948 and 1964 was a significant element of the federal government's sci-tech activities for a number of reasons. It was one of the first R&D programs in information storage and retrieval that involved a number of agencies. The office drew upon individuals from many fields and involved federal agencies and outside organizations in the planning and execution of its program. The initial success of its experimental projects in the use of mechanical and electronic devices encouraged other agencies to initiate experiments with these new technologies and routines. And finally, the Patent Office originated one of the earliest international cooperative R&D programs in information science under the aegis of ICIREPAT. After 1965, the Patent Office's R&D program was reduced in size and concentrated on adopting techniques and machines that had been developed by other organizations. This change was largely due to the lack of administrative and financial support within the Office and the Commerce Department.

REPORT OF THE BAKER PANEL OF THE PRESIDENT'S SCIENCE ADVISORY COMMITTEE

While the findings of the PSAC panel[5] headed by W. O. Baker, vice president

Federal R&D Programs and Studies 101

for Research of Bell Laboratories, are discussed in Chapter 4 and need not be repeated here, other aspects of this panel's activities warrant discussion. In its report to PSAC, the panel stated ". . . another area . . . also in need of great attention . . . has attracted little or no public interest. This is a matter of scientific information—the technical data a scientist needs in order to do his job." In addition, the panel took a strong position that U. S. efforts in sci-tech information activities should be decentralized, but better coordinated. The panel was even more specific in citing the need, ". . . to encourage and support a fundamental, long-term program of research and development, looking to the application of modern scientific knowledge to the overall problem through application of machine techniques and through yet undiscovered methods." This recommendation was the first governmental statement that R&D in the information sciences and technology should be expanded and that investigators from many fields of science should be encouraged to attack problems in this area. These vigorous statements would have had little impact on government policy and programs if additional actions had not been taken. After the President's Science Advisory Committee had endorsed the panel's position, the recommendations were presented at a cabinet meeting at which President Eisenhower asked each cabinet member in attendance if he agreed with the recommendations and would see to it that necessary action was taken in his department or agency. There was no dissent, and following the cabinet meeting, the top administrative officers of the departments and agencies assembled to discuss steps necessary to implement the recommendations. A White House press release of December 7, 1958, publicly announced this decision, and the next several years witnessed a rapid increase in federal support of R&D projects in information sciences and technology. The literature indicates no other such thorough, highly located consideration of a federal sci-tech information policy.

Other actions that stemmed from this policy determination included issuance in 1959 of Section 10 as an admendment to Executive Order 10521 that gave NSF responsibility for coordinating federal sci-tech information activities; establishment in 1958 of the Office of Science Information Service (OSIS) in the NSF; and formation that same year of the Federal Advisory Committee on Scientific Information (FACSI).

THE CRAWFORD TASK FORCE REPORT AND THE ROLE OF THE OFFICE OF SCIENCE AND TECHNOLOGY

Responding to a recommendation of the Crawford Task Force (see Chapter 4), the Office of Science and Technology (OST) assigned a full-time staff member to monitor and to coordinate federal sci-tech information and organized the Committee for Scientific Information (COSI) as a component of the Federal Council for Science and Technology (FCST). These actions enabled a representative of science information to present the need for improved information programs at meetings in the Executive Office of the President and at meetings of FCST and PSAC. As importantly, sci-tech information programs of the various agencies could be explained to the Bureau of the Budget. Also there was considerable interaction between White House specialists and Congress, which afforded another opportunity for agencies' sci-tech programs to be explained to members and committees of Congress. Later, this representation of sci-tech information activities in OST was strengthened when the secretary of COSATI was appointed and became a member of the OST staff. While accurate estimating of the value of this representation of sci-tech information programs at the level of the Executive Office of the President is not possible, certainly the growth of agencies' programs was aided by the efforts of the OST staff. An additional benefit of the monitorship of these OST staff members was their encouragement of cooperative projects among federal information agencies and suggestions for studies to assist in clarifying problems related to sci-tech information activities. Thus the Crawford Task Force Study indirectly aided federal information R&D programs as well as other activities.

WEINBERG REPORT TO THE PRESIDENT'S SCIENCE ADVISORY COMMITTEE

Another recommendation considered by the Crawford task force, but included in the recommendations made in 1963 by the PSAC panel headed by A. W. Weinberg, states, "Each federal agency concerned with science and technology must accept its responsibility for information activities in relevant fields, and must devote an appreciable fraction of its talent and resources to support information activities." Several federal departments and agencies expanded their internal programs, as well as their support of external sci-tech activities, as a result of this recommendation. Among these were the Department of Defense, the NIH and Public Health Service, the Department of Commerce, and NSF.

COSATI considered this recommendation at length and attempted to develop a scheme of allocating sci-tech information activities to appropriate departments and agencies. This concept was called "Delegated Agency." However, there were no clear-cut lines of demarcation between R&D programs of many of the agencies. For example, NASA, AEC, DOD, NIH, and NSF all were supporting research in biology; the same was true for chemistry. Thus, no agency had a unique mission to cover all of chemistry or biology. One attempt at coordinated support was made in 1964 when NSF, the Defense Department, and NIH agreed that NSF would assume leadership in strengthening the information services in chemistry and chemical engineering. At this time NSF embarked on a ten-year program of support of the American Chemical

Federal R&D Programs and Studies 103

Society's information systems development program. AEC explored a possible program with a similar objective with the American Institute of Physics but withdrew in favor of NSF in 1966. COSATI was never able to devise a comprehensive, acceptable allotment of responsibilities to implement the Delegated Agency concept, although the Weinberg report did stimulate increased support of R&D in information science, as well as improvement on on-going systems.

THE FRY AND HELLER STUDIES

Two studies funded by the National Science Foundation in 1962 and 1963 resulted in the collection and analysis of data that assisted materially in federal policy determination and program management.

The first was a study by George Fry and Associates titled *Survey of Reproduction of Copyrighted Materials.*[6] Both for-profit and not-for-profit publishers were concerned that the rapid development of copying devices might cause serious economic and other harm to the publishing industry. George Fry and Associates, after collecting and analyzing data on practices of copying from copyrighted materials, concluded:

> The results of this study, indicating that no significant economic damage occurs currently, must be viewed in the light of the following:
> 1. The development of new methods of information storage and retrieval could cause major changes. However, future changes are unpredictable at this time primarily because equipment and methods are in the "drawing board" stage.
> 2. Improved facsimile copying equipment would substantially alter the basic economics of the situation, and cause users to reevaluate their copying practices.

Within a few years, the second of the Fry qualifications became a reality. Among both users and publishers the concensus in 1969 was that copying of text, as well as the storage of the published material in computer-based systems, needed a legal resolution that would protect the rights of both users and publishers.

A number of mechanisms have been developed for payment of some kind of royalty fee for facsimile copying or for use of copyrighted material in a computer system but to date none of these have been widely accepted. Several publishers offer license arrangements for incorporating their bibliographic and other files in computer systems, but no mechanism has been generally accepted for payment for facsimile copying. The new copyright law will provide only a partial solution to these problems since no law can be drafted than can anticipate all the variant uses. Because publishers of copyrighted material have the largest economic stake in solving the copying and computer use problems, they should develop a royalty scheme to enable users to determine appropriate fees for copying or using material in a computer system. The private sector has been able to develop successful royalty systems in the field of the performing arts; it would seem reasonable that, with the increasing and widespread need for facsimile copies of or computer system access to scholarly material, a similar mechanism could be devised in this situation. It is this author's belief that the federal government should not be the operator and arbitrator of such a system since it also uses this type of material in its R&D programs.

The Heller Report

In 1963 the National Federation of Science Abstracting and Indexing Services issued a report prepared by Robert Heller and Associates entitled "National Plan for Abstracting and Indexing (A&I) Services."[7] This study was undertaken at the urging of, and was funded by, the NSF. The Heller Study produced extensive data on A&I activities in the United States, and its conclusions and recommendations created interest in both the federal and private sectors. The Heller investigators concentrated on economic and organizational problems—an approach that overlooked or ignored many technical problems facing these services. The study also underestimated the time and effort necessary to introduce new technology that was essential to achieve the principal recommendation in its report. The following conclusions are quoted from the Heller report:

> The basic elements of a sound abstracting and indexing system are established and in operation. However, work done on this assignment makes it clear that some of these elements require strengthening and that a number of basic changes are necessary to meet requirements in the future. Principal conclusions that are fundamental to development of a national plan are these:
> 1. The scientific community needs better secondary publications of both profession and project type.
>   a. Profession-type publications vary widely in format, coverage, and timeliness for various fields of science, with many below acceptable standards.
>   b. Many new publications of the project type are needed, since there is growing demand for interdisciplinary and more specialized intradisciplinary secondary publications.
> 2. Markets for the two types of publications vary widely.
>   a. Profession-type publications have little growth potential. The market is primarily restricted to institutional customers and is largely saturated.
>   b. Project-type publications have a large growth potential. Need for new products will expand as expenditures for scientific

research and development increase. Furthermore, since this type of publication is often designed for individual subscriptions, market expansion can be related to growth in number of scientists and engineers.

3. Forecasts of the growth of worldwide scientific literature coupled with a long-term upward trend of unit costs point to a significant problem. It is estimated that the cost of producing the proposed minimum publications in all fields of science would have been about $7 million to $9 million in 1961. Because of the growth of literature and rising unit costs, expense to provide the basic product will approach $15 million to $18 million in 1966 and $32 million to $38 million by 1971.

Initial steps in this work will require financial support from the National Science Foundation. It is recommended that grants to services be made for the purpose of implementing the program.

Recommendation 2: Establish an Operating Unit to Provide Project-Type Products

Many new project-type publications are needed and the profession-oriented services have much of the raw material necessary to produce them. It is recommended that those services establish on a joint-venture basis a central operating unit to utilize this raw material and produce new project-type products.

A schematic diagram of this new unit's relationship to the present abstracting and indexing system is shown on the next page. For purposes of identification in this report, the proposed unit is called "Organization X" [see Figure 6-2].

It should be emphasized that the new products of Organization X are to be determined by the requirements of the market. Publication form, indexing needs, timeliness, coverage and content—whether abstracts or citations—will vary widely depending on the market served. Correspondingly, the operations performed by the central unit may differ for each new product.

The technical and other advances made by the A&I services over the next few years enabled them to implement some of the Heller recommendations. Federal agencies were continually negotiating with the discipline-oriented services for use of parts or all of their data bases in the government's computer-based systems. In addition, several private organizations negotiated license agreements with the A&I services for the use of their data bases in centralized on-line services to institutions and individuals throughout the United States and Canada. The Canadian National Library of Science is acting as a kind of Organization X (see Figure 6-2) in offering services on the data bases of many of the U. S. abstracting and indexing services.

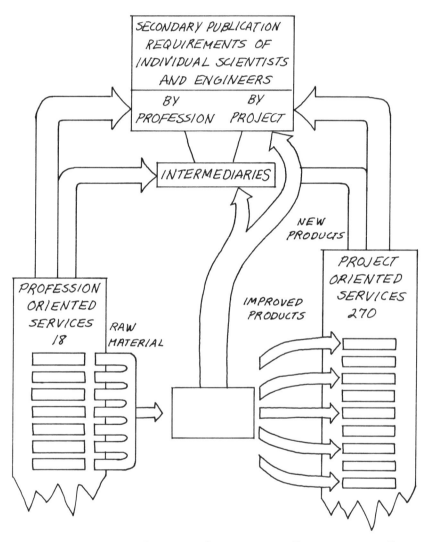

Figure 6-2. Schematic Diagram of Organization X. From a report by Robert Heller and Associates, Inc., in National Federation of Science Abstracting and Indexing Services, *A National Plan for Science Abstracting and Indexing Services* (Washington, D. C.: National Federation of Science Abstracting and Indexing Services, 1963)

THE NATIONAL SYSTEMS TASK FORCE

A major effort by the federal government to plan and initiate actions that would effect a coordinated and cooperative national sci-tech information system was that made by the COSATI Ad Hoc Task Group on National Systems. This task group's major contribution was the studies it commissioned to provide background for its planning and recommendations (see Chapter 4). Among these studies were one on abstracting and indexing by the Systems Development Corporation, another on a national documents handling system by the same corporation, a third on informal transfer of information by the American Institute for Research, and a fourth on data handling activities by Science Communications, Inc.

The recommendations of this Task Group were presented to the Federal Council for Science and Technology (FCST) in September 1965. The members of FCST were under pressure at that time to justify both to the Bureau of the Budget and to members of Congress the need for the continued increase in cost of R&D in science and technology. Questions were raised about the effectiveness of R&D efforts to solve the social and economic problems that were facing the nation. Since the task group's recommendations required both changes in legislation to clarify agencies' responsibilities in the sci-tech information field and increases in funds and staff to implement the program, FCST took no action on the recommendations. The plan was far too ambitious to be accepted by either the Executive Office of the President or committees of Congress. Despite the useful studies that were conducted by the investigating groups employed by the task group, the effort had little impact on the federal government's sci-tech programs.

UNISIST

Discussion of the work leading to establishment of the international program UNISIST is included here because it has involved many U. S. citizens and has the potential of materially affecting this nation's sci-tech information program.

In 1967 Unesco and the International Council of Scientific Unions (ICSU) agreed to cooperate in studying the feasibility of a World Information System—an activity that was but one component of a larger cooperative effort. To this end, they established a Central Committee and named as its chairman Harrison Brown of California Institute of Technology, who was also foreign secretary of the U. S. National Academy of Sciences. Brown was able to enlist the cooperation of a large number of individuals and organizations. Adam Wysocki represented Unesco in this program and was the administrative officer who implemented the studies commissioned by the Central Committee. Brown encouraged participation of representatives of many nongovernmental and international governmental organizations in the plenary sessions of the Central Committee. Scott Adams of the United States acted as chairman of an advisory panel and technical advisor to the committee. Jean-Claude Gardin of France was selected to write the report for the Central Committee, and a number of working groups were formed to make special studies that were used by the Central Committee and Gardin in the preparation of the final recommendations.

By 1969 the Central Committee's progress was sufficient to encourage Unesco and ICSU to agree that a conference should be held in 1971 to consider the committee's recommendations. In 1970 the Unesco General Conference approved holding an intergovernmental conference, October 4-8, 1971, in Paris. This conference was to consider establishing a world information system that would be called UNISIST. (This name for the system was adopted by the Central Committee; it is a meaningless acronym.) The basic concept of the world system was concisely stated by Brown in his letter of transmittal that accompanied the report to ICSU and Unesco; he wrote: "I am pleased to inform you that the committee agrees that a world science information system, considered as a flexible network of existing and future services, is feasible."

The committee's report[8] included a statement of five program objectives for UNISIST and 22 recommendations organized into the following six groups:

Tools of systems interconnections;

Effectiveness of information services;

Responsibilities of professional groups;

Institutional environment;

International assistance to developing countries;

Organization of UNISIST.

In addition, the report contained chapters on implementation of UNISIST, program priorities, and benefits and values of UNISIST.

The conference attracted eighty-two delegations from member countries of Unesco, ten from organizations in the UN system, and twenty from nongovernmental international organizations. There were long, heated debates over many of the twenty-three items in the final resolutions adopted by the conference.[9] The following excerpts contain the major recommendations of the conference (the numbers are those given the items in the final report):

7. *Supporting* the proposal made by the Unesco-ICSU Central Committee for the establishment of a UNISIST programme to advance a World Scientific and Technical Information System comprising a flexible network of existing and future information services.

10. *Inviting* the Director-General of Unesco to make adequate budgetary

Federal R&D Programs and Studies   109

provision to enable Unesco to play its leading role in the rapid establishment of the first stages of UNISIST, taking into account the needs of developing countries, providing necessary funds in the Programme for 1973-1974, as well as in the long-term outline plan of the Organization.
11. *Calls* upon other agencies of the family of the United Nations, and on other international organizations, both intergovernmental and nongovernmental, including those responsible for technical assistance to developing countries, to lend their full support and cooperation in the implementation of these recommendations.
12. *Requests* the Director-General of Unesco to convene periodically an Intergovernmental Conference to approve a long-term plan for UNISIST, and to review and evaluate the progress of the UNISIST programme.
12. *Recommends* that a steering Committee of 18 to 23 members be elected at the General Conference of Unesco from among its member states; this committee will supervise and, when necessary, revise the priorities of the programme within the framework of the long-term plan of action approved by the General Conference of Unesco and will report to the Intergovernmental Conference;
14. *Request's* the Director-General of Unesco, in consultation with ICSU and other organizations active in appropriate fields, to establish an Advisory Committee of scientists, engineers and information specialists reflecting the interests of both producers and users and those responsible for information transfer to assess periodically the ability of UNISIST programme to meet the needs of, and provide services to, the world's communities of scientists, engineers, and technologists, and to report its findings to the Director-General and to give advice to the Steering Committee as deemed necessary.
15. *Recommends* the creation within the Secretariat of Unesco of a unit of scientific and technical information which would act as a permanent secretariat of UNISIST and which would be responsible for the preparation and implementation of measures concerning the creation and the development of UNISIST.
16. *Recommends* that, initially, the above proposed unit take measures for the compatibility of existing and future national, regional, and international information systems;
17. *Recommends* further that at the same time special attention be paid to the complex and urgent needs of the developing countries and in particular their need for scientific and technical, as well as economic and social, information, for training (notably by scholarship programmes), and for provision of adequate infrastructure, and for stimulating or initiating new systems when needed.

Among the actions resulting from the deliberation of the Central Committee and the subsequent recommendations of the Intergovernmental Conference were:

1. The major abstracting and indexing services were assisted in their programs to develop standards for bibliographic descriptions and were able to work on standards for recording and transferring information on magnetic tape.
2. A World Serials Data Office was established in Paris with Unesco and French government support to coordinate and monitor a world system of numbering serials. Each country was to arrange for a national center to work with the Paris center. Several countries have established such centers, including the United States, which designated the Library of Congress.
3. Unesco has implemented the organizational structure recommended by the International Conference and appropriated funds for program support.

The reader may question the discussion of this large, complex international undertaking in a chapter on U. S. R&D programs and studies. The Unesco program, however, has been a catalyst for a number of U. S. actions. For example, the serials numbering data program has been furthered by this international program. NSF's Office of Science Information Service has changed its international program from concentration on nongovernmental organization activities to cooperation with governmental and intergovernmental organizations. Other federal agencies with international programs have become involved in the UNISIST activities. The impact of the program became more pronounced on U. S. sci-tech activities after 1972.

CONCLUSION

R&D activities in the United States have been vigorous and imaginative and together with other components of sci-tech services and activities have been under almost continual review by governmental study groups. This monitorship has been costly in time and money, but has surely led to changes that improved U. S. sci-tech information services, which are the most powerful in the world.

In the author's opinion, the decentralized approach in the United States has allowed the U. S. services to maintain world leadership. Even the USSR representatives recognized the need for flexibility in the sci-tech information field when they endorsed this flexibility concept in the UNISIST program. It is true also that there is need for a stronger coordinating mechanism in the United States—one that will have to be at a level above an individual agency. However, it seems rather fruitless to attempt a strong coordinated program for the governmental sci-tech information services as long as the R&D programs in science and technology are not organized into a coordinated national program. The information activities are services for the R&D programs and for the most part supported by them. It is rather difficult to coordinate services that are largely controlled and funded by an uncoordinated parent program. The one avenue of coordination that promises rapid and effective result is that of adopting common standards and techniques. Adoption of these will allow the transfer of information from one system to another. Federal support through the means of funding and coordination will go far to strengthen this important aspect of the sci-tech services in the United States.

NOTES AND REFERENCES

1. The members of the Interagency Group on Research on Information Systems included: Helen Brownson, Richard See, and Eugene Pronko of NSF; Gordon Goldstein, Marshall Yovitz, and Richard Wilcox of the Office of Naval Research; Al DeLucia and Z. Pankoweicz of Rome Air Development Center; Harold Wooster and Rowena Swanson of the Air Force Office of Scientific Research; Selig Starr and Fred Frishman of the Army Research Office; and Robert Taylor and Ivan Stherland of the Advanced Research Projects Agency, Department of Defense. Verner Clapp, Melvin Ruggles, and Laurence Heilprin of the Council of Library Resources were frequent participants.
2. National Science Foundation, *Current Research and Development in Scientific Documentation*, 15 vols (Washington, D. C.: Government Printing Office, 1957-1969). Volume 12 is an index to earlier volumes.
3. Harold Wooster, "Current Research and Development in Scientific Documentation," *Encyclopedia of Library and Information Science*, Vol. 6 (New York: Marcel Dekker, Inc., 1971), pp. 336-65. Wooster identified many of the investigators and notes their change in location from volume to volume.
4. Richard A. Spencer, "History of R&D in the Patent Office" (Washington, D. C.: U. S. Patent Office, 1972), 36 pp. (multilith).
5. The panel membership besides Baker included Curtis Benjamin, president of McGraw-Hill Book Company; Caryl P. Haskins, president of Carnegie Institution of Washington, D. C.: Elmer Hutchisson director of the American Institute of Physics; Warren C. Johnson, dean of the Division of Physical Sciences, University of Chicago; Don K. Price, dean of the School of Public Administration of Harvard University; and Allan T. Waterman and H. Scoville from the federal government.
6. George Fry & Associates, *Survey of Copyrighted Material Reproduction Practices in Scientific and Technical Fields* (Chicago: George Fry & Associates, 1962), out of print.
7. National Federation of Science Abstracting and Indexing Services, *A National Plan for Science Abstracting and Indexing Services* (Washington,

D. C.: National Federation of Science Abstracting and Indexing Services, 1963), 38 pp., prepared by Robert Heller & Associates, Inc., Cleveland, Ohio. Available from the National Federation of Abstracting and Indexing Services, Philadelphia.
8. Unesco, *The Synopsis of the Feasibility Study on a World Science Information System* (Paris: Unesco, 1971), 92 pp., a report by the ICSU/Unesco Central Committee, Unesco Doc. No. SC. 70/D. 74/A. A complete version of the Central Committee's report was issued at the same time by Unesco; it consisted of 161 pages.
9. Unesco, *Intergovernmental Conference for the Establishment of a World Science Information System (Paris, 4-8 October 1971), Final Report* (Paris: Unesco, 1971), 60 pp., Unesco Doc. No. SC/MD/25.

Chapter 7

Relationship between Federal Agencies and Nonfederal Organizations

The federal government has long been involved with nonfederal organizations in its sci-tech information activities. Chapter 2 describes a number of the very early cooperative efforts, perhaps the earliest of which was the purchase of Thomas Jefferson's library as the base for rebuilding the collections of the Library of Congress in 1815. A second example was the use of the National Institute for the Promotion of Science as a custodian for a short period of some of the federal government's specimen collections and the papers of Smithsonian between 1840 and 1846.

The first large-scale federal support of nonfederal institutions that involved sci-tech information was funding of the land-grant colleges and agricultural experiment stations that was authorized under the Morrill Act (1862) and the Hatch Act (1887). Within these institutions were built many of the nation's great research libraries, including those at the University of Wisconsin, University of Minnesota, University of Illinois, University of Colorado, and University of Washington. The widespread and effective dissemination of knowledge about advanced agricultural techniques resulted in part from the experiments, publications, and personal advice of the agriculture experiment stations and their staffs.

Another illustration of this federal relationship has been the encouragement and assistance of professional and scientific societies. The federal government worked closely with the American Association of Geologists (AAG) between 1840 and 1848 and since then with the American Association for the Advancement of Science (AAAS), which was the successor organization to the AAG. The American Geographical Society (AGS) was used by the Woodrow Wilson administration for studies (known as "The Inquiry") that provided important background for negotiating the peace treaty after World War I. The advice and support of the National Academy of Science (NAS), the American Association for the Advancement of Science (AAAS), the American Physical Society (APS), the American Chemical Society (ACS), and the American Institute of Electrical

Engineers (AIEE) assisted in the passage of the act that established the National Bureau of Standards (NBS) in 1901.

The distribution of catalog cards and U. S. government publications have been valuable to private libraries, research organizations, and individuals. The Government Printing Office (GPO) and other agencies have made their publications readily available to scientists and students, as have such science-oriented agencies as the U. S. Geological Survey, Coast and Geodetic Survey, Weather Bureau, the National Bureau of Standards, the Department of Agriculture, the Navy's Hydrographic Office, and the U. S. Army Corps of Engineers.

When technical and scientific reports became an important medium of dissemination, federal agencies established programs of distribution to regional libraries in the United States. This separate distribution was necessary because these reports were largely produced and distributed outside the GPO's normal publication and distribution program. Such agencies as the National Advisory Committee for Aeronautics, the Atomic Energy Commission, and the Office of Technical Services established regional report centers in U. S. public and research libraries. These agencies provided regional centers with both the actual reports and the cataloging, abstracting, and indexing information on them.

Five principal topics are discussed in this chapter. The first section is devoted to a review of federal agencies' efforts to extend their sci-tech information programs to private institutions, which they deemed necessary to further their R&D activities. Next is a discussion of the use of advisory panels and committees. Third is a review of the federal agencies' use of outside organizations to develop studies, planning systems, and programs and to conduct experiments in sci-tech information handling. The fourth section considers the federal government's assistance to other organizations to upgrade or initiate new sci-tech information services, including fostering the establishment of standards in library-information sciences by cooperating with nonfederal institutions. Finally, the role of the National Academy of Science/National Research Council is examined, since this organization was established to assist the federal government in science and technology.

THE EXPANSION OF FEDERAL AGENCIES' SCI-TECH INFORMATION PROGRAMS

The federal sci-tech agencies employed a variety of mechanisms to expand their R&D programs into different areas of the country. This dispersion of facilities was the result of many factors. In some cases, a particular geographical location was necessary for safety or because of the special resources available in the area. Thus the research laboratories of the Department of Agriculture were located where attention to special regional problems was needed. The Wood Products Research Laboratory was located in Madison, Wisconsin, in 1910, at the time that the Northwest Great Lakes areas were dominant in lumbering and forestry. The U. S. Geological Survey established a center at Denver, Colorado, because of the need for support of its geologists and surveyors working in western United States. The Bureau of Marine Fisheries in the Department of Commerce placed laboratories in key coastal locations such as Sandy Hook, New Jersey, and Woods Hole, Massachusetts. The Manhattan District Project located some of its laboratories in New Mexico and Tennessee for security and safety reasons. In some cases, the selection of a site for a R&D facility from among several good locations was based on political considerations. Almost as soon as a large federal R&D center was established, a sci-tech information service became a support unit in the facility. The following descriptions of a few of the federal sci-tech expansion activities illustrate the variety of mechanisms that the federal agencies employed to extend sci-tech information programs to the private sector.

The Atomic Energy Commission, when it assumed responsibility for the nuclear energy program in 1946, inherited the Manhattan District Project's manpower and laboratory resources. It contracted with a variety of organizations to operate its R&D facilities. In some cases, the laboratory was under the management of a private organization, for example, the Union Carbide and the General Electric Corporations. In other instances, a consortium of universities was organized to manage a laboratory as in the case of the Argonne National Laboratory near Chicago. The University of California operates the Lawrence Radiation Laboratory as well as other facilities for the AEC. Each of these widely dispersed centers was funded by the AEC to operate a sci-tech information center. Later the AEC contracted with established universities and private research libraries to be regional centers for nuclear science information. Among these were the libraries of the Georgia Institute of Technology in Atlanta, Georgia; the Massachusetts Institute of Technology in Cambridge, Massachusetts; the California Institute of Technology in Pasadena, California; and the John Crerar Library in Chicago, Illinois. As noted in Chapter 3, AEC also contracted with private institutions for the preparation, publication, and sale of manuscripts and with Microcard Corporation for the production of microcards of all AEC reports.

Associated with the research facilities of the National Advisory Committee for Aeronautics (NACA) had been a scientific library that operated with little cooperation or coordination with the libraries of the other NACA units until 1945. In that year NACA initiated a scientific reports program, in lieu of using only the scientific journals for dissemination of its scientific and technical findings. Soon thereafter, NACA developed a system of regional depositories of its publications in large research libraries such as the Library of Congress, the University of Illinois, and the University of Washington. In 1967, after the National Aeronautics and Space Administration (NASA) expanded and coordinated this program when it assumed the aeronautical and space research and exploration responsibility from NACA, it started a different type of technical information activity—the Technology Utilization Center. NASA contracted with private institutions and large universities to develop a service that would assist U. S. industry to utilize the new findings from its aero-space R&D programs. These centers were expected to initiate activities that would encourage industrial companies in identifying aero-space inventions and techniques that could be employed in producing new or better consumer products. One condition of this NASA contract was that the Technological Utilization Centers were to charge for their information and services in the hope that in a few years they would become self-supporting. NASA has also contracted for two other information services from nonfederal organizations. The first is an arrangement made in 1961 with Documentation Incorporated to establish a sci-tech information processing center to prepare, disseminate, catalog, abstract, and index information on federal reports and publications of interest to NASA. Periodically NASA has issued invitations for bids to operate this information processing center, and since 1972 it has been operated by Informatics, Inc. The second contractual arrangement is with the American Institute of Aeronautics and Astronautics (AIAA) to acquire, catalog, index, and abstract the world's published literature in its field, other than that produced by the U. S. government. The AIAA prepare its materials according to NASA specifications so that the catalogs, indexes, and abstracts can be merged with those prepared at the NASA technical information processing center. The AIAA also publishes an index and abstract bulletin covering the published material; NASA issues a counterpart for government documents.

Another approach to extending an agency's sci-tech information service to the private sector is exemplified by an agreement between the Department of Agriculture Library and the University of California at Davis. The Davis Campus Library became the department's depository and service center for catalogs on agricultural machinery. Similar arrangements were made by the Library of Congress when it turned over its collection of foreign dissertations and these to the Midwest Interlibrary Center in Chicago for custody and service. In extending information services to its field employees, Department of Agriculture also contracted with the University of Nebraska Library to furnish library services to department employees working in that region. This same arrangement has been made in other areas where the department of Agriculture does not have its own sci-tech information facilities. The regional medical library service that the National Library of Medicine organized in response to the Medical Library Assistance Act of 1965, which represents still another method employed by a federal agency to extend its library and information services, is described in Chapter 5. Prior to that time, predecessors of the National Library of Medicine arranged, at different times, with various contractors for such services, including the Carnegie Corporation of Washington, and the American Medical Association for publication of *Index Medicus.*

The National Science Foundation contracted with Herner and Company to prepare, distribute, and collect questionnaires from all potential investigators in library-information sciences. The contract also called for Herner and Company to analyze and organize the information obtained from these questionnaires for publication as an issue of *Current Research and Development in Scientific Documentation* (CRDSD). Other NSF agreements were made with scientific societies to select, translate, and publish Soviet and other important foreign sci-tech materials. NSF agreed to pay the difference between cost and income of these projects. The societies were given varying numbers of years, usually three to five, to make these translation publications self supporting.

In 1968 the Bureau of Public Roads contracted with the Highway Research Board of the National Academy of Science/National Research Council to develop and operate a Highway Research Information Center. This center furnishes services to both state and federal highway research projects. It also maintains close liaison with other national and international highway information centers.

The agencies of the Department of Defense also extended their information activities by establishing information centers. These Information Analysis Centers, as they were later called, collect and evaluate scientific and engineering data needed in the Defense Department R&D programs. Examples are the Metals and Ceramics Information Center of the Battelle Memorial Institute, the Chemical Propulsion Information Agency at the Johns Hopkins Applied Physics Laboratory, and the Electronics Properties Information Center at Purdue University. In 1970 the Department of Defense changed its policy for support of these centers by requiring them to charge for their services. (Several other agencies such as NASA, AEC, EPA, and NSF sponsor similar centers that support their own as well as other R&D programs.[1])

A variant of the above program is that of the National Standard Data Reference Service (NSDRS) of the National Bureau of Standards. It was authorized by Congress to be the U. S. focal point for evaluation of scientific data for R&D. The NSDRS has working arrangements with a number of private organizations in the United States, under which they evaluate, organize, and disseminate critical data under international rules established by the Committee for Data for Science and Technology (CODATA), an international coordinating office that is sponsored by the International Council of Scientific Unions (ICSU).[2] A number of the Information Analysis Centers established by the Defense Department are part of NSDRS's system, and NSDRS represents the United States on the governing board of CODATA.

In sum, federal agencies have employed a variety of mechanisms to extend and expand their sci-tech information products and services through nonfederal organizations, as illustrated by the following activities:

1. Support of sci-tech information activities in contractor laboratories;
2. Designate nonfederal libraries as information centers and furnish these with reports and the bibliographic tools for giving service;

3. Establish centers to disseminate information on scientific and technical innovations;
4. Support university and other private laboratories to maintain information analysis centers specializing in critical (evaluated) numerical data;
5. Use of university or other research libraries to store, organize, and give service on specialized publications;
6. Contract with nonfederal libraries to be regional service centers for sci-tech reports
7. Contract with private organizations to perform technical information processing services or maintain sci-tech information facilities;
8. Contract with private organizations to produce special sci-tech publications such as directories, translations, indexes, and monographs.

ADVISORY PANELS AND STUDY GROUPS

Almost every federal scientific and technical information agency relied upon individuals and organizations outside the government to investigate its operations or to advise on its programs and policies. The AEC had a Technical Information Panel and an Advisory Group of Librarians to assist in the development of its programs and to advise on services and products. The National Science Foundation, under a mandate in Title IX of the National Defense Education Act of 1958, organized a Science Information Council to advise and review the program of the Office of Science Information Service. In 1948, the Library of Congress established a task group to aid in the preparation of a statement on its future role; in 1961, the Library, with funds from the Council on Library Resources, organized another group to study "Automation and the Library of Congress"; and in 1963 this group, which was chaired by Gilbert King, issued a report under the above title. Other agencies had continuing or short-term advisory or study groups to aid with policy and operations problems.

An interesting use of outside talent was made by the Department of Defense when it initiated Project Lex in 1968. This project, administered by the Office of Naval Research, was directed to prepare a subject heading list and index thesaurus for use by the Defense Documentation Center and other Defense Department information agencies. DOD invited scientific societies and federal and nonfederal organizations to appoint individuals to work with the Project Lex staff in the preparation of this thesaurus.

The federal government has and continues to use many scientists and other specialists to provide advice on a large variety of sci-tech information problems. These range from advice on a particular technique or operational problem to suggestions for a major change in policy or development of a large, complex information system. The methods for obtaining this advice vary from small ad hoc groups to formal committees. Congress, in several instances, requested departments or agencies to seek outside advice and has passed legislation requiring such action.

Relationship between Federal Agencies and Nonfederal Organizations   119

The following are examples of the federal government's use of nonfederal experts to assist in improving information activities:

Short-term use of experts—NSF in 1958 convened a group of experts to review its support of mechanical translation R&D projects that resulted in NSF orienting its support towards projects in computational and other linquistic research.

Use of a committee to advise on an agency's systems development program—The Secretary of Commerce appointed in 1955 a committee to recommend ways to improve the Patent Office's information system that resulted in a joint National Bureau of Standards and Patent Office R&D program known as HAYSTAQ.

Use of the National Academy of Sciences—In 1966 the NSF and OST requested the NAS to form a committee to review and recommend a plan to improve U. S. sci-tech activities that resulted in the SATCOM report of 1969.

Continuing advisory committee authorized by Congress—The Science Information Council of NSF advised the NSF from 1958 to 1972 on its support program in science information.

SYSTEMS, PROGRAM DEVELOPMENT, AND OPERATIONAL EXPERIMENTS

Following World War II, the federal sci-tech information agencies increasingly incorporated new technology into their operational systems, and by 1958 several sci-tech information agencies were considering the development of complex information systems that would take advantage of computer technology, rapid facsimile reproduction, microforms for storage and distribution, and telecommunications. During the next seventeen years, these agencies increasingly supported operational experiments prior to adoption of such new techniques, which often required systems planning and development—rather than the piece-meal introduction. Since most agencies initially had little experience in this approach, they turned to the private sector for assistance.

About 1955, the Navy Department gave Xerox Corporation a several-million-dollar contract to develop a rapid facsimile reproduction system. The first effort was not entirely successful, but the Armed Services Technical Information Agency (ASTIA), the Library of Congress, and several other agencies indicated they would purchase or rent an improved model. The result was ASTIA's installation in 1957 of a new facsimile reproduction system, the basic component of which was the "Copyflow" unit. ASTIA began operation with seven Copyflow machines that were soon running twenty-four hours a day to keep up with ASTIA orders. The Library of Congress and other agencies almost immediately installed Xerox equipment. In 1971, the Defense Documentation Center (DDC) and the Clearinghouse for Federal Scientific and Technical Information (CFSTI) indicated that they would install an improved system being developed by Xerox, and in 1972 DDC and the National Technical Information Service (NTIS) signed rental contracts with Xerox for the 3600-Met System. Figure 7-1 shows Copyflow machines that use the electrostatic process for reproduction.

Figure 7-1. Copyflow machines in operation. Rolls of paper are fed through the machine where images are recorded; paper is then cut into page sizes. Courtesy of Defense Documentation Center

In 1958, the National Library of Medicine (NLM) let two contracts that initiated the systems development program for the Medical Literature Analysis and Retrieval System (MEDLARS) and the Graphic Arts Composing Equipment (GRACE). One contract was with Photon, Inc., for the development of a rapid, computer-controlled composition system that would handle multiple fonts of type. The second contract was with the Information System Division of the General Electric Company for design and development of a comprehensive medical literature system. The evidence indicates that MEDLARS and GRACE constitute one of the first comprehensive systems development programs initiated by an information agency of the federal government. In 1964, NLM instituted its Demand Search Service using MEDLARS, and in 1965 it began to issue its Recurring Bibliographies, such as the *Cerebrovascular Bibliography*. Three years later the NLM began preliminary development of the second generation of MEDLARS by letting a contract to Auerbach Corporation to develop the specifications for an invitation to bid on the design of the advanced system.

In 1958, ASTIA began to explore possible use of a computer in its information activities. In 1959, it let a contract to Remington-Rand Corporation for the installation of a SS90 Univac system. (IBM did the preliminary study for ASTIA that led to the invitation to bid on which Remington-Rand was successful.) Four years later, Remington-Rand installed a Univac 1107 system for ASTIA which had a 500 million alpha-numerical character drum storage component.

During the interval between 1960 and 1968, many federal information agencies initiated systems programs, with most agencies drawing upon the experience of private organizations for designing, and often for development of these systems. For example, the Library of Congress turned to United Aircraft Corporation, NASA to Bunker Ramo Woolridge and Lockheed Corporations, and the National Agricultural Library to Booz-Allen, Inc., and EDUCOM. Many other illustrations could be cited.

In sum, the examples of arrangements between the federal information agencies and private, largely for-profit organizations in the development of large automated information systems attests to the vitality and effectiveness that can be achieved in government-private sector cooperation. In a fifteen-year period— 1957 to 1972—the federal information agencies by cooperation with the private sector organizations had changed their manually operated information systems, which were limited in speed, currency, and diversity of products, into operating systems capable of producing a wide variety of products and services, and rapidly becoming available to remote users through communications networks. It appears likely that even greater developments will occur during the next decade through advances in computer, satellite, and photographic technologies.

FEDERAL ASSISTANCE TO ORGANIZATIONS IN SCI-TECH INFORMATION ACTIVITIES

The federal government has long followed a policy of encouraging and frequently assisting scientific and professional societies and other organizations with their sci-tech information activities. For example, the Geological Survey has cooperated with state geological surveys in joint mapping and resource inventory programs. It has also supported field work by members of university faculties and published the results of this research as part of its professional paper series. The Office of Education has been authorized to support research and otherwise aid public libraries to improve their collections and services. In 1933, the National Bureau of Standards was authorized by Congress to pay page charges on articles published in scientific society journals, when an article has been written by an NBS scientist. In 1961, this policy of paying page charges for articles in sci-tech journals became government wide.

Sci-tech meetings sponsored by societies have had strong and continuing support from federal government agencies, with the aid being extended in numerous ways. Government agencies have made grants to support conferences and have furnished funds to sci-tech societies to assist in publication of conference proceedings, or the support has been in the form of purchase of a significant number of copies of the proceedings or as an unrestricted grant to assist in publication. Government agencies have also been generous in allowing their scientists and technical personnel to participate in societies' planning and conducting national and international meetings and in allowing staff attendance at and participation in such conferences. In other cases, federal agencies have either cosponsored meetings or contracted with societies to plan and conduct small work sessions on topics of particular concern to the agency.

This cooperation and support has not been purely altruistic; the federal agencies are anxious to have their sci-tech personnel maintain their professional stature, and many of the meetings result in direct benefits to agencies' programs. No accurate data are available on the number of federal employees who participate in such meetings. Some measure of the federal government's financial contribution is indicated by the expenditure in 1972 of $21.3 million for sci-tech meetings and symposia. Although this amount includes funds that federal agencies expended for meetings they planned and conducted, the greater portion of it went for the support of conferences held by sci-tech societies.

Another category is direct or indirect support of sci-tech journals and monographs. An example already mentioned is the honoring of page charges by societies for articles published as a result of research conducted with federal funds. Accurately determining the annual expenditure of federal money for this purpose has not been possible, but an estimate made in 1964 by NSF's Office of Science Information Services placed it at over two million dollars for that year. Direct subsidy of sci-tech journals of professional societies has not been a general federal practice, although several federal agencies have, from time to time, aided societies to publish backlogs of accepted papers and to prepare cumulative indexes to journals. In 1962, NSF made grants totalling $830,000 in support of primary publications. After that year, NSF began to reduce its support to primary publication in view of the federal government's policy of honoring page charges as part of costs of research grants and contracts.

Other federal agency techniques for supporting publications have been funding monographic series with the understanding that the publishing organization would return funds in excess of the costs of publication, agreeing to purchase sufficient copies to pay for the initial publication costs, agreeing to pay the deficit for publication of translated foreign journals incurred by a society and allowing a grantees to pay for publication of monographs that resulted from research projects.

The support of society, discipline-oriented abstracting and indexing services has been a major area of federal agencies involvement and support that has been extended in various ways. The Geological Survey was one of the first scientific agencies to cooperate with a professional society in the preparation of an index to publications in the geological sciences. This agency in 1880 began to issue a bibliography of geology of North America. In 1934, when the Geological Society of America (GSA) began a *Bibliography and Index to Geology Other Than in North America*, the Geological Survey assisted by making office space available for its staff and allowing them to use the Geological Survey library and map collection as source material for the bibliography. This cooperative agreement continued until 1969 when the Geological Survey, GSA, and the American Geological Institute (AGI) agreed jointly to publish a single bibliography and index. The AGI (with funds from NSF until 1974) prepares the entries; GSA assumes responsibility for publication and sale of the *Bibliography and Index to Geology*; and the Geological Survey continues to provide library and map collection resources and now assists the AGI through purchase of special bibliographies and by making annual grants.

Another cooperative arrangement was the agreement of the Department of Agriculture Library in 1946 to accept requests from members of the American Chemical Society and subscribers to *Chemical Abstracts* for a copy of any article cited in *Chemical Abstracts*. This arrangement continued until 1956 when the practice was challenged by publishers as an infringement of copyright. *Meteorological Abstracts* provides another example of federal agency assistance in establishing and maintaining an abstracting and indexing service. In 1950, the Air Force contracted with the American Meteorological Society to start an abstracting service in meteorology, climatology, and allied subjects. The Weather Bureau cooperated in this effort by allowing its library resources to be used by the Meteorological Society and by assigning staff to assist in the selection and abstracting of publications. When the Air Force withdrew its financial support after a few years, NSF partially funded the production of *Meteorological Abstracts*, with the Weather Bureau continuing its cooperative assistance.

From 1958 to 1971 the major funding source for discipline-oriented abstracting and indexing services was NSF's Office of Science Information Service. Between 1958 and 1964, OSIS aided many A&I services by financing deficits, underwriting the cost of initiation of new products, supporting experiments on new automated techniques, and partially funding the National Federation of Science Abstracting and Indexing Services (now National Federation of Abstracting and Indexing Services). OSIS soon realized that this ad hoc support for individual projects was not encouraging a coordinated program either within the services or among the several services. At the NFSAIS annual meeting in 1961, OSIS encouraged the Federation to prepare a national plan, which resulted in the Robert Heller and Associates recommendations issued in 1963, to which the A&I services of the various societies were not in a position to respond affirmatively (see Chapter 6).

In 1964, OSIS alerted the A&I services that it was unwilling to continue the program of supporting individual projects and insisted that each service that wished support first prepare a systems plan and indicate how each project would aid in the development of the overall systems program. Since the A&I services had neither the funds nor the special talent needed to develop systems programs, the NSF agreed that it would fund systems studies by the societies.

In 1965 the American Chemical Society submitted an acceptable five-year systems plan. It indicated this was the first phase of a longer development program. The five-year proposal was funded by NSF with some financial assistance from the Department Defense and NIH. In succeeding years, the American Institute of Physics, the American Psychological Association, and the American Geological Institute submitted plans prepared with NSF support, while other societies accepted funding from NSF to develop systems programs but did not submit proposals for a systems development program. (A detailed analysis of the society systems programs is the subject of Berninger's doctoral dissertation.[3]) The societies that initiated systems programs in the late 1960s had vastly improved operations by 1972, and by 1974 had developed fiscal structures that allowed them to operate without significant federal subsidy.

In addition to encouraging the systems development programs, NSF encouraged the A&I services to explore ways in which their products could be merged to meet the increasingly specialized needs of mission-oriented programs. For example, NSF supported experimental projects that investigated methods of searching several files for references to a common topic. Two of these were projects at the University of Georgia and at the Illinois Institute of Technology. The NSF also assisted in funding the work of the American National Standards Institute's Z39 Committee on Standards for bibliographic elements and encouraged the societies to participate in this program. Another project partially supported by NSF was a joint effort by the Chemical Abstracts Services, Biological Abstracts, and Engineering Index to develop common standards and techniques that would facilitate their exchanging information and apportioning coverage of journals that all three services cover. A similar effort by the International Council of Scientific Unions Abstracting Board was also encouraged by NSF through support of the board and provision of travel expenses for American representatives to attend ICU/AB working group meetings.

In sum, the federal government agencies' policy of providing encouragement and financial support to discipline-oriented services was an important factor in enabling abstracting and indexing services in the following fields to achieve or maintain preeminence among the world's A&I services: chemistry, astronomy, mathematics, earth sciences, biology, meteorology, aerospace, metals, engineering, psychology, physics, and many specialized fields.

NATIONAL ACADEMY OF SCIENCES

The National Academy of Sciences (NAS) is unique among private organizations in the United States. It was established in 1863 by a congressional act of incorporation, which said in part, ". . . the academy shall, whenever called upon by any department of the government, investigate, examine, experiment, and report on any subject of science or art . . . ." The enabling act also indicated that expenses for projects requested by the federal agencies should be paid from these agencies' appropriations, but the NAS was not to be otherwise compensated—that is, it was to receive no direct federal appropriations.

Congress clearly intended that NAS should provide a mechanism by which federal agencies, including Congress, could obtain scientific and engineering information and advice. However, the scientists who were the original members of NAS had a different concept of the organization. They viewed its role to be that of a prestigious scientific establishment in many ways similar to the Royal Society of London or the French Academy of Sciences. History has shown that the Academy has fulfilled both roles.

Prior to and during World War I, NAS was requested by the federal government to investigate and advise on many sci-tech problems. To respond effectively to these requests in 1916, NAS established the National Research Council (NRC) as its operating arm. The NRC was officially approved by the federal government when President Wilson issued an executive order in May 1918. The National Academy of Engineering (NAE) was chartered by Congress in 1966, and in 1968 NAS established the Institute of Medical Sciences (IMS). These two organizations also use the NRC's operating capabilities.

Because Congress decreed that NAS was to be reimbursed by the government only for costs related to requested projects, it turned to private sources for funds to maintain its continuing operations and for construction and maintenance of buildings and other facilities. A major source of funds for endowment as well as support of NAS/NRC activities has been the private foundations such as the Carnegie Corporation, the Rockefeller Foundation, and the Ford Foundation. Many other organizations and individuals also have contributed funds in support of NAS/NRC, and the same is true of NAE and IMS.

Since a complete review of the NAS/NRC-NAE science information activities would be too voluminous for inclusion in this book, the following account is limited to a few illustrative activities. One of the first large science information activities of the NAS/NRC grew out of a NRC committee appointed in 1922 to review problems related to the evaluating, organization, and dissemination of standard data on varified physical and chemical properties of substances (so-called critical data). As a result, a project known as the International Critical Tables was organized with support from federal agencies, private foundations, industrial organizations, and universities. This project collected and evaluated critical data that met designated standards from both governmental and non-governmental organizations, both in the United States and abroad. The Mc-Graw-Hill Book Company issued a seven-volume publication entitled *International Critical Tables of Numerical Data 1926-1930.* No attempt has been made since to publish so comprehensive a compendium of critical tables.

In 1957 the NAS/NRC established the Office of Critical Tables to continue acting as a catalyst and coordinator of projects for evaluating and disseminating critical data. In 1968 the role of this office was enlarged to include other data of interest to both scientists and engineers, and various NRC committees have given attention to data standards for nomenclature, symbols, and units in such fields as nuclear energy, oceanography, physics, chemistry, biology and geology.

NAS/NRC has responded to requests from government agencies and other organizations to act as a sponsor or a catalyst in initiating new information tools or services. For example, in 1921 the NRC Division of Biology and Agriculture was asked to arrange a meeting of representatives of societies and other organizations interested in forming a federation of biological societies and in improving the information resources in biology. (This session, held as part of the American Association for Advancement of Science (AAAS) annual meeting in Toronto, Canada, in December 1921, was followed by a series of meetings, studies, and surveys conducted to identify the interests and needs of the biologists, a detailed account of which will be part of a forthcoming *History of Biological Abstracts* to be published by *Biological Abstracts.* The result of these activities was the formation of a Union of American Biological Societies, funded with a grant from the Rockefeller Foundation that also provided support for the development and operation of abstracting services in biology. The final result in 1926 was the start of *Biological Abstracts* by a consolidation of several smaller efforts. J. R. Schramm executive secretary of the NRC's Division of Biology and Agriculture from 1922 to 1924, played an important role in planning and promoting the new publication and was the first editor of it. Although this first issue indicates that *Biological Abstracts* was published under the sponsorship of the Union of American Biological Societies, the effective participation of the NAS/NRC catalyzed the action that resulted in the birth of *Biological Abstracts,* which continues to be the major English-language abstracting and indexing service in the biological sciences.

The NAS/NRC was also instrumental in the catalyzing action that resulted in the initiation of the *Bibliography of Economic Geology* in 1929 under the auspices of the Society of Economic Geologists. Later in the 1960s, NAS representatives participated in the deliberations that resulted in starting the American Geological Institute's Geo-Ref Service. Other fields in which NAS/NRC has aided in initiating information tools are electronics, Pacific Islands ecology, geology, atmospheric sciences, building science, fire research, energy research, and medical sciences.

The NAS/NRC has made significant contributions to science and engineering by accepting responsibility for organizing and operating specialized information services. In 1920, the NAS/NRC established the Highway Research Board at the request of the federal government and with the concurrence of state highway commissions and private organizations. The board has encouraged experiments on road building materials and highway construction. In line with these responsibilities, the Highway Research Board collected and organized pertinent publications, data, and other information. By 1965, these information resources had become so voluminous that the board decided to establish a specialized information center, known as the Highway Research Information Center, with Paul Irick as the director; it became operational in 1967. The center has become an important focal point for information on highway research and has provided effective liaison with similar centers in various states and European countries.

In 1946, at the request of the military agencies, a committee was formed to investigate problems related to the effects of chemicals on biological functions. This group recommended that an information center on chemical-biological coordination be founded to facilitate R&D in this area. This Chemical-Biological Coordination Center operated from 1946 to 1957 when the Defense Department withdrew its support. This center and its NAS/NRC monitoring committee made two especially important contributions. First was the development of a classification code for organizing information in this area; this code was published with NSF support in 1960. Second, the center was among the first groups to experiment with organizing and searching large, complex information files with a punchcard system.

As described in Chapter 3, federal agencies interested in developing an inventory of current research projects in the medical sciences persuaded the NAS/NRC to establish the Medical Science Information Exchange, which began operations in 1950. The project proved so successful that the participating agencies asked that biological and psychological research projects be included in the inventory. Because of administrative problems, however, the exchange was transferred to the Smithsonian Institution in 1953 and the name changed to the Biological Sciences Information Exchange.

When planning for the International Geophysical Year (IGY) (July 1957 to December 1958), federal agencies and other participating organizations were aware of the need for centers to collect, organize, and disseminate data and other information that would result from the geophysical research projects. Organizations in other participating countries were aware of the same need. Under the sponsorship of the International Council of Scientific Unions, an agreement was reached on the establishment of three World Data Centers—one in USSR, one in Europe, and one in the United States. The NAS/NRC was selected to organize and coordinate the U. S. efforts, with support from NSF. World Data Center A, as the U. S. activity was named, arranged for specialized subcenters in universities, federal agencies, and private organizations. The center also acted as the liaison link with the other World Data Centers. Because of the continuing need for these services, World Data Center A and many of the specialized centers have continued to operate even though federal support has been drastically reduced. In addition, the NRC-IGY Committee prepared the 301-page *United States IGY Bibliography, 1953-1960*, which was published in 1963 as NAS/NRC publication 1087.

The NAS/NRC has also responded to many requests to sponsor meetings to consider information problems in specific disciplines or subdisciplines. In 1958 it departed from this pattern by cooperating with the American Documentation Institute and the NSF in organizing and conducting an international conference to consider all aspects of sci-tech information. Scientists, librarians, and information specialists from all parts of the world attended this conference. Prior to the meeting in November 1958, the organizing committee had arranged for papers covering a wide variety of subjects. The conference was significant in that representatives from many countries were subsidized so that they could attend and participate; it was the first large international meeting on scientific information following World War II; and the organizing committee was able to encourage participation from many fields of science and engineering, from libraries, and from investigators working on information problems and developing new information-handling techniques.

In 1965 the NRC National Committee for the International Federation for Documentation (FID) hosted the Federation's annual meeting in Washington, D. C. This conference differed from previous FID congresses in that it provided for symposia and sessions of contributed papers that could be attended by all interested individuals. Usually an FID congress has been limited to meetings arranged by FID Committees and Working Groups.

In recent years, NAS/NRC has extended its long history of participation in international scientific conferences, committees, and working groups by naming individuals to represent it at these sessions to include scientific information conferences. Among the first of these was a conference in England in 1948 that was sponsored by Unesco but organized by and held at the Royal Society in London. The topics for consideration were publications, and abstracting, and indexing of scientific material. One result of this meeting was the formation of the Abstracting Board of the International Council of Scientific Unions (ICSU-AB). Also described earlier was the participation of NAS/NRC representatives on international committees for standardization of nomenclature, symbols, and units.

Another task NAS/NRC undertook in response to requests from federal agencies has been performance of studies and reviews of problems in the sci-tech information field. NRC committees have investigated modern methods of handling chemical information, systems of linear notations of chemical compounds, status of research in mechanical translation and computational linguistics, automation in libraries, copyright problems raised by new techniques for rapid facsimile reproduction and computer systems with on-line capabilities for handling information, and methods of handling medical information. One of the latest (1965-1968) NRC study committees, called SATCOM (Committee on Scientific and Technical Communications), was established at the requests of NSF Foundation and the Office of Science and Technology and was one of the first NRC committees sponsored by both NAS and NAE. The committee's report in both complete and synopsis versions has been widely distributed and vigorously discussed. Entitled *Scientific and Technical Communication: A Pressing National Problem and Recommendations for Its Solution*, it suggests ways to improve sci-tech communications in the United States.[4] Although no positive actions can be directly attributed to this committee's recommendations, the hearings it held and the discussions about them drew the attention of congressmen, federal administrators, industrial executives, and university administrators to urgent problems in sci-tech communication.

The above study and report was followed in 1972 by another done by the Information Systems Panel of the Computer Science and Engineering Board. The report, *Libraries and Information Technology: A National Systems Challenge*, was done at the request of the Council on Library Resources.[5]

From the time of its establishment, the NAS has been a publisher of important scientific and technical monographs, bulletins, and reports. Its publications number in the thousands. The latest bibliography (1965) contains titles that indicate very wide scientific and engineering coverage.

In 1959, the NAS/NRC was persuaded by NSF to establish an Office of Documentation, with the following responsibilities to provide liaison and coordination among the many NRC committees and other groups who consider sci-tech information problems as part of a larger sci-tech; to aid in identifying sci-tech problems in the U. S. nonfederal community on which federal action should be taken; and to act as a liaison office with national and international organizations having sci-tech information programs and activities. The NAS/NRC organized a small staff and appointed a committee to advise this new office on its program and activities. Although the Office of Documentation has been partially successful in all three areas of its assigned responsibilities, both NSF and NRC were apparently disappointed by its lack of effectiveness. When NSF withdrew its support in 1968, the director of the office resigned; while it has continued to operate with a reduced staff, its NRC advisory committee has ceased to function.

In sum, although NAS has never considered sci-tech information to be one of its priority responsibilities, it has played an important role in the development and conduct of information programs in the United States. The above review of NAS/NRC-NAE activities suggest its contributions can be identified as follows: maintaining committees or offices involved in scientific information programs, organizing and operating information facilities in response to federal agencies requests, acting as a sponsor or catalysts in establishing new information services, designating participants in information conferences and participating in the conduct of such conferences, investigating and recommending on aspects of U. S. sci-tech information programs when requested by federal agencies, issuing publications that contribute to sci-tech information, and finally, organizing and operating the Office of Documentation.

NOTES AND REFERENCES

1. Committee on Scientific and Technical Information, *Directory of Information Analysis Centers* (Washington, D. C.: Federal Council for Science and Technology, Committee on Scientific and Technical Information, 1970), 71 pp.
2. R. B. Gavert, R. L. Moore, and J. H. Westbrook, *Critical Surveys of Data Sources: Mechanical Properties of Metals* (Washington, D. C.: U. S. Department of Commerce, National Bureau of Standards, 1974), 81 pp., National Bureau of Standards Special Pub. 396-1, printed by the Government Printing Office. This is the first title in a series to be sponsored by the Office of Standard Reference Data, National Bureau of Standards.
3. Douglas E. Berninger, "Strategic Planning and Decision Making in a Federally Funded Scientific Information Program," Unpublished Ph.D. dissertation American University, Washington, D. C., 1975. Available from University Microfilms, Inc., Pub. No. 75-19656.
4. National Academy of Sciences, *Scientific and Technical Communication, A Pressing National Problem and Recommendations for Its Solution* (Washington, D. C.: National Academy of Sciences, 1969), 336 pp., NAS Pub. 1707. This report was also issued as a 30-page synopsis.
5. Computer Science and Engineering Board, *Libraries and Information Technology: A National Systems Challenge* (Washington, D. C.: National Academy of Sciences 1972), 84 pp.

Chapter 8

The Federal Government and Foreign Exchange of Information

Scientists have always been anxious to exchange information and ideas with their peers in other countries. National governments, except in times of emergencies, have fostered this exchange. In the new American republic, the commissioner of patents began to exchange patents with other countries soon after his office was established in 1836. In 1840 Congress recognized the need for exchange of documents by authorizing the Librarian of Congress to exchange duplicate copies in the Library's collections. This exchange program was expanded in 1866 and 1867 when Congress passed resolutions authorizing the congressional printer to make copies of each printed document available for use by the Library of Congress for exchange purposes. In the early 1850s, Congress sanctioned the exchange program of the Smithsonian Institution for scientific publications (see Chapter 2) by authorizing and funding the Office of International Exchange, which is the unit in the Smithsonian that receives, collects, and transmits the publications. Beginning at an early date, ambassadors and other foreign service officials, such as the "plant explorers" of the Department of Agriculture, have actively participated in the acquisition of foreign publications, and in 1866 the U. S. minister in Brussels, Belgium, agreed with representatives of seven other countries on a "Convention for International Exchange of Documents, Scientific and Literary Publications." This convention, which was approved by Congress, has been revised and expanded numerous times, and since then, many other governmental conventions, treaties, and agreements to foster the international exchange of information have been established. In addition, every federal agency that requires foreign publications and other information has developed its own exchange agreements with the appropriate foreign agencies and private institutions. Today there are literally thousands of exchange agreements between U. S. federal agencies and foreign institutions.

In the National Science Foundation enabling legislation of 1950, Congress included a clause authorizing the NSF to "foster the interchange of information among the scientists in the United States and other countries." This interchange of sci-tech information at the government level has not been based solely on exchange of publications. The federal government has supported international meetings of scientists and engineers; exchanged bibliographic information and methods for organizing, storing, and disseminating sci-tech information; negotiated agreements for exchange of scientific and technical data; participated in international programs, such as the International Geophysical Year, for the collection, organization, dissemination, and analysis of data; promoted international standards by enacting legislation to allow for importation and exportation of biological, physical, and anthropological specimens and scientific equipment; and fostered and frequently supported programs to enable U. S. scientists and engineers to work with their peers in laboratories in both the United States and abroad.

COOPERATIVE EXCHANGE PROJECTS WITH FOREIGN ORGANIZATIONS

Following World War II, the declassification and dissemination programs of the Publications Board and subsequently the Department of Commerce were oriented primarily toward releasing important foreign and domestic sci-tech information to the U. S. sci-tech community. These efforts did make it possible, however, for the Office of Technical Services (OTS) to furnish sci-tech information first to European countries under the Marshall Plan and later to the developing countries under contracts with the Agency for International Development. The Atomic Energy Commission (AEC) had its own declassification program that at first benefitted only U. S. sci-tech community and later (1955) was extended to other countries.

In 1958 Congress passed PL-480, a section of which authorized the use of U. S. foreign surplus credits for translating and procuring publications, as well as for supporting research in other countries. Two programs were established as a result of this legislation. The first was an interdepartmental translations program that was coordinated by the Office of Science Information Service (OSIS) of the National Science Foundation (NSF). Using funds appropriated under this act, the OSIS established translation projects in Israel, India, Poland, Yugoslavia, Tunisia, and Burma. In most cases federal agencies identified the publications they wished translated; occasionally the translation center would suggest titles for consideration. Each participating U. S. agency was given an annual allocation of pages it could request to be translated.

Between 1959, when the program got underway, and 1972 several thousand pages of sci-tech material were translated and printed by these centers. Although each agency could make distribution of the material it commissioned, OTS was designated as the public outlet for all of the translations. The emphasis was on

Soviet sci-tech publications, but some translations were made from other languages, including German, and those of Eastern Europe. An unexpected benefit of this program was the opportunity for some of the translation centers to become acquainted with the authors of the works being translated. Some authors suggested a later publication or agreed to revise and update material scheduled for translation. This program has continued since 1972, although reduction of U. S. surplus foreign credits in countries with competent scientific translators has caused some parts of it to diminish or to be phased out. The Israeli program terminated in 1973 but the Israeli Center for Scientific Translation had sufficiently expanded its program between 1960 and 1973 that it continues to operate without subsidy from the U. S. foreign credits.

The second program under the PL-480 authorization was initiated by the Library of Congress. Using both regular appropriations and the U. S. surplus credits, the Library established a number of acquisitions and cataloging centers. Locations of some centers are in South America, the Middle East, India, and Israel. Personnel from the Library of Congress administer these centers and give technical guidance to the staff which is largely local. These PL-480 centers have allowed the Library to strengthen its acquisitions and cataloging programs and, in addition, it has been able to furnish copies of publications to U. S. research libraries.

Selection of centers by both the Library of Congress and NSF were restricted to countries where the United States had a surplus of foreign credits. In addition, the participating countries had to agree that the projects could be maintained with local currency. These nations benefitted in two ways. First, the centers provided both employment and training for local citizens. Second, the PL-480 funds provided the means to develop competent centers that would not have been possible without the cooperation and technical guidance of the U. S. agencies.

Other federal agencies, such as the Public Health Service, the Smithsonian Institution, and the Departments of Agriculture and Interior, took advantage of this authorization to have scientific and technical papers prepared, bibliographies compiled, and state-of-the-art analyses made. Larger research programs sponsored by a number of federal agencies have also been supported by surplus foreign credits.

In addition to these international programs, the large federal sci-tech information centers have developed exchange agreements with their foreign partners. For example, the AEC Technical Information Section developed exchange programs with Canada, the United Kingdom, Belgium, and France, among others. When the European countries organized the European Atomic Energy Agency (EURATOM), AEC extended its information exchange program to include EURATOM. The agreement included exchange of bibliographies, abstracts, publications and so forth; however, it also contained provision for the exchange of information on techniques, equipment, and systems. This latter activity resulted in some exchange of personnel that permitted each agency to become familiar with the other's techniques and procedures.

When the International Atomic Energy Agency (IAEA) was established in 1957, one of its major programs was to foster the exchange of information among its members. This program led to an agreement in 1966 to develop an international information agency that was the International Nuclear Information Services (INIS). The participating countries agreed that their bibliographic information on nuclear energy would be available to INIS in a prescribed format and that INIS would organize the various inputs and make the consolidated data available to the national centers. INIS would also respond to inquiries from national nuclear energy centers which were not equipped to exploit the INIS bibliographic data file. As of 1972, this international cooperative program was still in the developmental stage.

The National Aeronautics and Space Agency (NASA) took a somewhat different approach to its international information exchange program. In addition to the bilateral exchange of publications, which is the basic element of most exchange agreements, NASA developed a close working relationship with the European Space Research Organization (ESRO). This agreement entailed ESRO's collecting publications and preparing the bibliographic record in a prescribed format that could be used in the NASA system. In return, NASA assisted ESRO in developing a counterpart of NASA's information system for use in the European space research community. In 1969, ESRO, with NASA cooperation, installed its RECON which is an online, interactive, computer-based system for remote access to central files.

The above programs exemplify international information exchanges that go far beyond the earlier agreements based solely on exchange of documents. Other examples are the cooperative arrangements the National Library of Medicine has with other countries, such as the United Kingdom, Australia, and France, as well as with international organization such as the World Health Organization and the Pan American Health Organization. The National Agricultural Library participates in the World Food and Agriculture Organization's program to develop a network for document and bibliographic information transfers. U. S. federal agencies and private organizations are cooperating in the development of the International Standard Serials Numbering System (ISSN) sponsored by the UNISIST component of Unesco. This venture is dedicated to making information on serial titles more readily available. The International Standard Book Numbering System (ISBN) is another cooperative program in which U. S. organizations are participating. The Library of Congress also has a number of bilateral agreements on the use of its MARC files of cataloging information on tape.

CODATA is an international program that grew out of scientists and engineers interest in exchanging standard data for research and development. The U. S. data activities have two focal points for coordination: one is the National

The Federal Government and Foreign Exchange of Information   135

Academy of Sciences/National Research Council (NAS/NRC) with its Office of Critical Tables and advisory committee, and the other is the federal government's National Standard Reference Data Service (NSRDS) in the National Bureau of Standards (see Chapter 7). The United Kingdom designated the Office of Scientific and Technical Information (OSTI) as its coordinating agency, and the USSR established the State Service for Standard Reference Data (GSSSD) with broad responsibilities for data evaluation and dissemination. The principal mission of CODATA is to encourage the production and distribution of critically evaluated data. One of its first projects was the preparation and publication of a volume listing active compilation projects throughout the world.[1] Concurrently, CODATA formed task groups in a number of special areas which included computer handling of numerical data, key values of thermodynamic properties, chemical kinetics and fundamental constants, presentation of data in primary literature, and data accessibility. CODATA illustrates an international program consisting of many projects in individual countries that are coordinated through a national program with national representatives, in turn, working cooperatively through a nongovernmental international organization, International Council of Scientific Unions (ICSU). In 1972, thirteen countries had representatives on CODATA. In addition to the cooperative program with CODATA, U. S. private organizations and federal agencies have many bilateral arrangements with organizations in other countries for exchange of standard data.

The International Geophysical Year (IGY) illustrates a different approach to collecting, organizing, and disseminating numerical data. In this international project, sponsored by ICSU, countries made plans for coordinated geophysical observations during the eighteen months from July 1957 to December 1958. In addition, data collecting and disseminating centers (World Data Centers) were organized, and bibliographies of publications resulting from this worldwide endeavor were published (see Chapter 7).

National and international cooperation has been essential to the development of the worldwide system for collecting and disseminating data related to weather and climate. In the middle of the nineteenth century, scientists interested in meteorology met informally in Europe to agree on standards for making observations and for recording and disseminating weather data. These meetings resulted in the formation of the International Meteorological Organization (IMO) in Rome, Italy, in 1879. Within this organization was an International Meteorological Committee (IMC) made up of the directors of the national weather services. The IMO and IMC did much to standardize the collection, organization, and dissemination of data affecting weather and climate. The story of the work of the IMO is recorded in Henrik Gerrit Cannegeiter's history,[2] and the U. S. efforts are described in Patrick Hughes' work.[3] The following comments are limited to two examples of international cooperation.

The first U. S. weather satellite (TIROS) became operational in 1960. In 1966, the federal government organized the National Operational Meteorological

System that has used both conventional and satellite data to make its predictions. With satellite data and pictures becoming available, in 1968 the World Meteorological Organization (WMO) mounted a project called the World Weather Watch. World Watch Centers were established in Melbourne, Australia; Moscow, USSR; and Washington, D. C. Each center is maintained by its national government. The centers agreed to perform the following functions:

1. Receive conventional and satellite data on a global basis;
2. Transmit these data to other world, national, and regional centers;
3. Prepare meteorological analyses and prognoses;
4. Make these analyses and prognoses available to other centers;
5. Act as international training centers;
6. Conduct programs of meteorological research;
7. Act as a data archives for international use.

Although the centers were committed only to perform the above functions until 1971, they have continued to operate as international centers.

The second international weather program resulted from a U. S. technical innovation that became operational in 1971. In that year, Automatic Picture Transmission (APT) became available internationally. Through this system, satellite pictures of weather for use in weather and storm forecasting can be received by five hundred stations in fifty countries. By receipt and interpretation of these pictures and other satellite data, a station located remote from an incipient storm can identify the storm and alert places in its path. For example, stations in the United States have been able to alert India, the Caribbean countries, and other areas to impending storms that were developing over remote ocean regions, but were headed for these populated areas.

While these two programs are the more striking examples of weather-related international cooperation, other international programs of great importance have been for the worldwide publication and dissemination of weather records, weather maps, and treatises on meteorological research.

THE FEDERAL GOVERNMENT AND NONGOVERNMENTAL ORGANIZATIONS

As noted in Chapter 7, meetings of working groups and committees constitute one of the most effective ways in which scientists and engineers exchange information. The federal government has participated in such international programs described in the previous section as those fostered by the International Council of Scientific Unions and the World Meteorological Organization and has also furthered programs of nongovernmental international organizations by granting funds, for example, to pay dues of their U. S. members. Further, federal agencies frequently make grants to pay travel expenses of scientists participating in nongovernment international conferences or committee meetings, and also to send federal scientists to the conferences and meetings where they have been selected as participants. Another practice of the federal agencies has been assisting a U. S. organization financially and to host nongovernment conferences or committee meetings, and at times federal agencies have determined that a proposed nongovernment international project will further its own program, and has thus provided support for the project. Federal agencies have also frequently participated in nongovernment international conferences by placing exhibits at these meetings. (See earlier chapters for specific examples of these activities, particularly the discussion of NAS/NRC in Chapter 7.)

SCI-TECH INFORMATION PROGRAMS OF OTHER COUNTRIES

Several countries have been developing sci-tech information systems of particular interest to the United States. Three of these are discussed briefly in the following sections.

USSR

The USSR has been building large sci-tech information services for more than twenty years. These have been centrally planned, with program and policy carefully controlled at the national level. The operations, however, have been quite decentralized. *The USSR Scientific and Technical Information System: A U. S. View*,[4] prepared in 1973 by a team of U. S. experts that visited the USSR, summarizes the activities of the Soviet services. The unit of the Soviet government that is responsible for national planning, policy, and monitoring of sci-tech information is the Directorate for Scientific and Technical Information and Its Dissemination, of the Senate Committee for Science and Technology, Council of Ministers. N. B. Arutiunov heads the directorate. Twelve Moscow-based all-union institutes and agencies provide centralized resources for support of the National system. Nine of these, including the All-Union Institute of Scientific and Technical Information (VINITI), are under direct policy and program direction of the Directorate for Scientific and Technical Information and Its Dissemination although they may be part of another Soviet agency. For example, VINITI is administered as an Institute of the USSR Academy of Sciences. Three all-union institutes are influenced, but not closely controlled, by the directorate. They are the Central Institute of Scientific Information on Construction and Architecture, the All-Union Research Institute for Medical and Medical-Technical Information, and the All-Union Institute for Scientific and Technical Information on Agriculture.

The following excerpts from the U. S. experts' report, referred to above, describe the structure and interrelationship of the Soviet sci-tech information system. In addition to the directorate and the all-union institutes mentioned above, there are . . .

> 1. Eighty-two branch or industrial information networks, corresponding to the industrial ministries comprising the Council of Ministers, USSR. Each is headed by a Central Institute. INFORMELECTRO . . . is the Central Institute for the Ministry of Electrotechnology.
> 2. Fifteen Institutes of Scientific and Technical Information serving the 15 Union Republics . . . .
> 3. Seventy-two territorial "Interbranch" information institutes. These are subordinate to the Republic Institutes in the larger Republics . . . .
> 4. In addition, there are approximately 10,000 departments of scientific and technical information in research institutes and enterprises (Otdel Nuchnoi i Tekhnicheskoi Informatsii (ONTI) and some 23,000 scientific and technical libraries. The team of experts characterize the flow of information in the following manner: Information flows down, up, and horizontally within the Soviet industrial heirarchies. First, the All-Union Information agencies, such as VINITI and the State Public Library for Science and Technology, generate information materials which flow downward either directly or through the Central Institutes to the branch systems and the specialized and Territorial Interbranch Centers to the ultimate users in enterprises and scientific institutions. Unpublished information about scientific and technological innovations in enterprises and institutions moves upward to branch centers and to the All-Union Center of Scientific and Technical Information and the All-Union Exhibition of Economic Progress.

The visiting delegation also described in the following manner how the Soviet manpower needs in the sci-tech information field are met:

> There are three professional categories involving staffing the Soviet information system; scientist and engineers, systems analysts and computer experts, and librarians.
>
> To improve the qualifications of the profession groups for management and management positions in the information institutes, the State Committee had established in 1972 an Institute for Advanced Training of Information Officers.

This institute offers short courses—two months full time and five months part time—to scientists and engineers to orient them in information work.

The Soviets have at least two methods of measuring the value of information to the users. First, Selective Dissemination Service is discontinued if participants do not regularly supply feedback on evaluation forms that are provided to them. Second, the Soviets have a pricing system: One purpose is to identify user satisfaction through their willingness to purchase items; two other purposes are to recover the basic cost of the service and to stimulate the utilization of information materials. The team of experts noted, however, that many services were free.

The work of coordination and standardization is headed by three advisors to the State Committee who

> ... serve as members of a group to coordinate information systems development in different elements of the system with VINITI's proposed integrated information system. Another group is working on plans to standardize computer applications in the Union Republic Centers.

The Soviet system has had a marked influence on the information programs of other countries in the Soviet bloc. These ties were further strengthened after 1969 when the International Center of Scientific and Technical Information was established in Moscow. The center was organized under the auspices of the Council of Mutual Economic Assistance (CMEA), but it has its own governing board. Its mission is to achieve compatibility among the participating countries' national systems. Linguistic and technical standards are high on the list of priority projects being undertaken by the international center. Although the international center was established by eight Eastern European countries and the People's Republic of Mongolia, other countries may participate. The USSR is the major source of support for the center. It has provided the director, Yuri N. Sorokin, formerly of VINITI; it contributes 74 percent of the annual budget; and it has constructed a new headquarters building, which was completed in 1973. One of the center's early projects was the collection of information on sci-tech serials published by member countries. It is expected that the international center will become the agent for the participating countries in the Informational Serials Date System being developed as a component of the Unesco-UNISIST program. Another center project is the development of a chemical information system to serve participating countries.

Although there have been few formal information exchange programs between the United States and USSR, officials from information agencies and other organizations have visited each other's institutions and actively explored possible joint ventures. The first formal visit of Soviet information specialists to the U. S. was in 1958 when a Soviet delegation participated in the International Conference on Scientific Information. The latest U. S. formal visit in June 1973 resulted in the report that is quoted above. The Soviets returned this visit in the fall of 1973.

Canada

In 1965, Canadian authorities began to review their sci-tech information activities and issued a series of reports. The most extensive was published in 1969 by the Science Council of Canada as Special Study No. 8.[5] This study and several others were used by the Science Council in preparing recommendations contained in its Report No. 6,[6] which is known as the Katz report. On the basis of these reports and recommendations, the Canadian government is extensively modernizing its sci-tech information system.

Although the Katz recommendations and the other studies stimulated increased Canadian government action, The National Science Library (prior to 1966, the Library of the Canadian Research Council) has been developing a cooperative sci-tech information program with other Canadian libraries and making extensive use of U. S. sci-tech information resources and services. As Jack E. Brown, director of the Canadian National Science Library (NSL) stated in his progress report[7] to the National Federation of Abstracting and Indexing Services in March 1972:

> In the past, Canadian librarians and others have been too content to rely on the information resources and services so abundant in the U. S. and made so readily available.... Traditional Canadian publications have been well covered by abstracting and indexing services produced in the U. S. A.

According to this progress report, by 1971 Canada's National Science Library program included:

1. A question-and-answer service provided by reference and research staff trained in sci-tech subjects, competent in foreign languages, and skilled in using the keys to the world's sci-tech literature;
2. Provision of loans and photocopies of materials not readily available in other parts of Canada;
3. A clearinghouse for technical reports and government publications issued by national and international agencies (These are normally reproduced as microfiche copies; the NSL has appropriate equipment to provide either copies in microfiche or enlargements of the pages on the microfiche);
4. Publication of a wide variety of directories, indexes, and other tools designed to facilitate the use of Canada's sci-tech literature resources;
5. A Translation Index and Information Center that identifies and locates translations of foreign language scientific and technical papers prepared in all parts of the world;
6. National selective dissemination of information through a service called

the CAN/SDI program, which is a computer-based, current-awareness service that alerts individual scientists, engineers, and others to the existence of recently published papers in their specific fields of interest.

Other significant elements of the Canadian system are the Health Science Resource Center, which is a unit of the NSL, and the National Research Council's Technical Information Service (TIS), which has field officers working directly with Canadian industries on their information problems. These officers are well acquainted with the NSL's resources and services.

The NSL has made extensive use of U. S. information resources on magnetic tape in the CAN/SDI program, and in 1972 the following U. S. tape services were being used: *Chemical Titles, C A Condensates, Biological Abstracts, ISI Source and Citation,* American Geological Institute's *Geo-Ref, NEDLARs, Pollution Information Projects* (PIP), *Compendex, Metadex,* and NTIS. Other Canadian and U. S. government working arrangements for exchange of information include for example, those between the U. S. and Canadian Defense Departments, the USGS and its Canadian partner, and AEC and its Canadian counterpart.

United Kingdom

In 1965, a new U. K. government agency, the Office for Scientific and Technical Information (OSTI), was established in the Ministry of Education and Science with Harry Hookway as the head. Formation of OSTI resulted from increasing concern in the United Kingdom over the problems of handling the rising tide of sci-tech information. In 1963, a small research unit had been set up by the former Department of Scientific and Industrial Research (DSIR) to investigate scientific information problems, and when OSTI was formed, a nucleus staff was transferred from DSIR to the Department of Education and Science to build the new program.[8] OSTI's responsibilities were to concentrate on R&D, education, and coordination in sci-tech information. It was not to become responsible for operating services unless no suitable organization existed. The intention was that OSTI would work closely with government and independent bodies responsible for libraries and information services.

To obtain advice on policy requirements, in 1965 the secretary of state of the United Kingdom set up an Advisory Committee for Scientific and Technical Information (ACSTI). This committee consisted of twelve independent members—mainly scientists, technologists, and administrators—and an equal number of representatives from government departments and agencies with strong interests in sci-tech information. ACSTI was concerned in particular with the work of OSTI and coordination of its activities with those of government departments and agencies and independent organizations. ACSTI also considered national libraries serving science and technology and annually reviewed the work of the National Lending Library for Science and Technology. In addition, it presented reports and recommendations to the National Libraries Committee.

Very early in its program, OSTI supported programs that foster cooperative projects with U. S. organizations. Among these were experiments with computerized data bases provided by Chemical Abstracts Services, Biological Abstracts, the National Library of Medicine, the Atomic Energy Commission, and the Institute for Scientific Information (ISI). OSTI made certain that methods for evaluating these experimental services were incorporated into the projects, and the results from these experiments have proved useful to the U. S. organizations involved.

OSTI also maintained liaison with national and international sci-tech information organizations. For example, frequent discussions were held between OSTI and OSIS regarding the experiments with U. S.-produced bibliographic data bases, some of which OSIS partially funded; staff members from the two offices were detailed for short periods to work in each other's programs; OSTI supported data centers for the British National Committee on Critical Data as part of that committee's contribution to CODATA; and the head of OSTI participated in the work of UNISIST and was chairman of the Information Policy Group of the Organization for Economic Cooperation and Development.

Within the United Kingdom, OSTI supported programs in education and research on information problems and systems development. To illustrate, OSTI supported an experiment to have university information officers aid faculty and graduate students with their information problems, made grants to investigators of science information problems, and aided the Institute of Electrical Engineers to develop a computerized system for the production of the following, *Science Abstracts* and *Information Services in Physics, Electrotechnology, Computers and Control* (INSPEC).

In 1972 Parliament passed the British Library Act that established a National Library under the control and management of a new board.[9] The British Library, as the act named the new organization, consists of the British Library Board of not less than eight members, supported by a staff headed by a secretary, Harry Hookway who was formerly Head of OSTI. Among the organizational units transferred to the control of the board are the Library of the British Museum OSTI, which has become the Research and Development Department of the British Library.

The board appoints advisory committees to represent the different sectors of activities under its purview. The purpose of the reorganization is to strengthen the U. K.'s information services through development of a strong national library system in which sci-tech information activities will be an important component.[10]

Traditionally, British and U. S. organizations have had extensive information exchange programs. For example, U. S. physicists rely on and partially support the physics abstracting incorporated in *Science Abstracts* produced by the

British Institute of Electrical Engineers. In turn, the British chemists use the U. S.—based *Chemical Abstracts* as their primary service. The British Iron and Steel Institute and the American Society for Metals (ASM) cooperate in the production of the ASM's *Metals Reviews*. The British, as well as several other countries, cooperate with the American Society of Automotive Engineers in the production of the latter's data handbook. In short, U. S. and British organizations have a considerable variety of information exchange agreements.

INTERNATIONAL GOVERNMENTAL ORGANIZATIONS

In many instances, effective international information exchange is best achieved through international organizations rather than through bilateral agreements because many nations participate in the sci-tech programs. In addition, when a scientific or technical effort requires significant funding to generate large amounts of data and other information, such as was true of the International Geophysical Year, the project is best placed under the guidance of an international governmental organization, since the cooperation of many governments is probably the only realistic source of the funding.

The list of IGOs participating in information exchange is long and the types of activities varied. Some of these organizations and their affiliates are the specialized agencies of the United Nations, including the Industrial Development Organization (UNIDO), which encourages the exchange of technical information among the developed countries and the less-developed countries; the World Meteorological Organization (WMO), which has fostered development of standards and dissemination techniques for meteorology and weather; the International Atomic Energy Agency (IAEA), which is developing an atomic energy information network (INIS); Unesco, with its assistance to nations to develop documentation centers, libraries, and archives, and its sponsorship of the development of a worldwide network of sci-tech information systems (UNISIST). Since its beginning as part of the Marshall Plan, the Organization for Economic Cooperation and Development (OECD) has had a large technical information component, and the Organization of American States has a program to improve libraries in member states and also sponsored projects to upgrade sci-tech information activities. Other IGOs and their programs, such as the International Center for Scientific and Technical Information in Moscow, the Committee for International Cooperation in Information Retrieval among Examining Patent Offices (ICIREPAT), the International Geophysical Year's World Data Centers, the International Catalogue of Scientific Literature, the World Weather Watch, and UNISIST, have all been mentioned in earlier discussion. Therefore, only one example is described in this section.

Early in the 1960s the OECD had an advisory group consisting of directors of National Documentation Centers. This advisory group performed a useful function in fostering the exchange of information on techniques and operational programs, but OECD was anxious to encourage the development of national policies and program plans for sci-tech information. In 1965, OECD organized an Information Policy Group (IPG). This group consisted of national representatives who had responsibility for governmental policy, planning, and coordination of sci-tech information activities. To obtain an overview of member nations' sci-tech information programs, the group encouraged OECD to include sci-tech information in its program of reviews of each member country's R&D activities. Each such review entailed a study of the member nation's R&D activities by a team of experts, preparation of a report of the team's findings for the member nation, discussion of the findings before the OECD national representatives by the team of experts and representatives of the member nation, and publication of both the study and the discussion. The IPG used these studies as well as others as an aid in preparing recommendations for the improvement of policy formulation and program planning for national sci-tech endeavors. While IPG was not successful in preparing a set of guidelines for national policy formulation and program planning, it did perform a number of useful functions by facilitating the exchange of experiences among the participating members, bringing to the attention of R&D administrators the need for a well planned sci-tech information program to support R&D, making suggestions for training and education of information specialists, identifying problems that hindered the development of a coordinated system of national sci-tech information activities, investigating the level of budgetary support required for a viable national sci-tech information program, and identifying elements needed for an effective R&D effort in information sciences for the continued improvement of a national sci-tech information system. Probably one of the IPG's greatest contributions was fostering bilateral cooperative projects between sci-tech information organizations. The IPG was involved in experimental projects in Europe that concerned use of the NLM and CAS tape services.

In sum, the reader should realize that most large international projects, including those fostered by IGOs, require mixed participation of governments and private organizations at both the national level and the international level. An illustration is the concept of an international standard serials numbering (ISSN) system: the coordinating office for this widespread effort is part of Unesco, an international governmental organization; the success of the system depends on national governmental agencies, which must give both financial and other support to coordinate their national efforts; private national organizations, which must place an ISSN number on each of their serial issues; and nongovernmental international organizations, which must encourage participation of their members in using the ISSN on their publications.

CONCLUSION

Exchange of information and ideas among scientists and engineers is the life blood of the advancement of science and technology. Since new scientific discoveries and new technical advances are not limited to one country, nations have encouraged and supported activities and organizations that foster this information exchange. The U. S. government has been no exception. As with domestic information activities, no single agency or international organization can be the focal point for sci-tech information exchange among nations. Therefore the U. S. federal government has cooperated with a wide spectrum of domestic, foreign, and international organizations in furthering information transfer between its scientists and engineers and those of other countries.

NOTES AND REFERENCES

1. See Committee on Data for Science and Technology of the International Council of Scientific Unions, *International Compendium of Numerical Data Projects—A Survey and Analysis* (Berlin: Springer Verlag, 1969).
2. Hendrick Gerret Cannegeiter, "The History of the International Meteorological Organization 1872-1951," *Annalen der Meteorologie*, Neue Folge Nr. 1 (Offenbach A M: Des Deutchen Watterdienstes, 1963), 279 pp.
3. Patrick Hughes, *A History of the Birth and Growth of the National Weather Service: 1870-1970* (New York: Gordon and Breach Publishers, Inc., 1970), 212 pp. This publication was written by a Weather Service official.
4. National Science Foundation, *The USSR Scientific and Technical Information System: A U. S. View* (Washington, D. C.: National Science Foundation, 1973), 63 pp., report of U. S. participants in the U. S./USSR Symposium on Scientific and Technical Information held in Moscow, June, 18-20, 1973, National Technical Information Service Report No. NTIS-SR-73-01.
5. Science Council of Canada, *Special Study No. 8, Scientific and Technical Information in Canada* (Ottowa: Queen's Printer, 1939). This special study is in two parts. Part I consists of recommendations; Part II gives the background and considerations on which Part I is based. Part II is divided into seven chapters published separately, and each chapter discusses a sector of Canadian sci-tech activities.
6. Science Council of Canada, *Report No. 6: A Policy for Scientific and Technical Information Dissemination* (Ottowa: Queen's Printer, 1969), 35 pp. This report is often referred to as "The Katz Report" after the chairman of the Science Council Committee on Scientific and Technical Information.
7. Jack E. Brown, *The Canadian Scientific and Technical Information System: A Progress Report* (Philadelphia: National Federation of Science Abstracting and Indexing Services, 1972), 26 pp. NFSAIS Report No. 4 (1972 Miles Conrad Lecture).
8. See Department of Education and Science, *OSTI: The First Five Years*

(London: Her Majesty's Stationery Office, 1971), 64 pp. This work was done by the office for Scientific and Technical Information, 1965-1970.
9. British Library Act of 1972; Chapter 54 (London: Her Majesty's Stationery Office, 1972), 10 pp.
10. *The British Library*, presented to Parliament by the Paymaster-General by Command of Her Majesty, January 1971 (London: Her Majesty's Stationery Office, 1972), 7 pp.

Chapter 9

Trends, Persons, and Future Directions

A review of federal sci-tech information operations, policies, and programs during the 1942-1972 period can be organized in several different ways, including the following:

1. The basic information services for R&D programs have always been and continue to be founded on collections of materials. These materials must be acquired, organized, and serviced by persons who are familiar with both library and information sciences and with the subject matter covered by the R&D program. During this period, established sci-tech libraries and sci-tech departments in large general libraries expanded markedly. In addition, several thousand specialized libraries were organized. These libraries either were a part of federal R&D programs or were in federally funded, independent R&D projects. The libraries specialized in service on a variety of materials. Thus, today there are federal library collections of aerial and ground photography, maps, engineering and other drawings, sci-tech reports, sci-tech data, as well as material in conventional monographic and serial formats. A number of these specialized libraries are units of larger, complex information activities, and many of the large sci-tech libraries incorporate information activities relatively new to them.

2. During and following World War II, many specialized information services were organized. One type was the sci-tech information center. Establishment of these centers was stimulated by the increased use of the sci-tech report as a medium to achieve currency and flexibility of dissemination of sci-tech information. Another type of information activity was the inventory of current R&D projects. These inventories were often developed in technical information centers. Such is the case in the Department of Defense where the Defense Documentation Center has responsibility for organizing and maintaining a centralized current-research inventory. The Smithsonian Science Information Exchange represents the interdepartmental approach. The third type of new information activity is the referral center. Such centers specialize in collecting information on "who is doing what, where." They are not intended to provide information on particular topics, but are able to refer the inquirer—whether individuals or organizations—to sources that probably can answer the substantive questions.

3. Data collection, organization, analysis, and dissemination projects expanded markedly during the 1942-1972 period. Laboratory and field observations were greatly accelerated through introduction of new instrumentation and techniques. These types of data required organization for ready dissemination to researchers for use in their R&D projects, and most of this information required further analysis before it could be relied upon. A second type was data that had been evaluated. This evaluation was performed in many places throughout the United States and other countries. Data information analysis centers were fostered by the federal government in both federal and nonfederal laboratories. The collection, organization and dissemination of these data became an increasingly important element of the sci-tech information sector during this period.

TRENDS

Growth both in size and complexity is the major characteristic of library and other information services during this period. For example, in 1940 the Library of Congress collections in science and technology numbered 1,500,000 volumes. By 1972, this number had grown to 4,500,000. This increase in quantity alone placed a heavy burden on the library to acquire, organize, and service these materials. This burden was complicated by the receipt of sci-tech materials in formats formerly very few in number or not represented at all. For example, increasing numbers of sci-tech reports were acquired by the Library. Material in microfilm or other facsimile reproductions began to arrive in large quantities by the end of World War II. Increased interest in sci-tech materials from foreign countries resulted in the Library's being flooded with publications in foreign languages, especially documents in Russian and other languages of Eastern Europe.

The story of the Library of Congress efforts to organize and give service on its rapidly growing and increasingly complex collections exemplifies efforts of many research libraries both within and outside the federal establishment. In 1942 the Library had one professional person specializing in securing and providing service on sci-tech documents. A number of events occurred between 1945 and 1955 that markedly changed the Library services. First, in 1945, was the influx of captured German documents and the release of publications held during the war by foreign organizations as part of their exchange agreements with the Library. The German documents were forwarded to the Library as part of the Commerce Department's Office of Technical Services' program to disseminate these materials quickly to U. S. industry. The Library's Photoduplication Service, established in 1938, reproduced these items for distribution. In addition, cataloging and other bibliographic records had to be prepared so that the documents could be incorporated into the Library collections. Many of them were on microfilm, airgraphs, or in other facsimile reproduced forms. As a result, the Library had to devise new methods for incorporating the materials into the collections and for giving service on them, which resulted in expansion and improvement of its photoduplication service as well as its cataloging of and other services on documents in microfilm or other reproduced forms.

The next major impact on the Library of Congress's services was the acceptance in 1947 of a contract from the Office of Naval Research to acquire, catalog, index and abstract, and give service on sci-tech reports of interest to the Navy's R&D programs. The librarian of Congress assigned the responsibility for this Science and Technical Project to Mortimer Taube, who had to recruit a staff with knowledge of and experience with sci-tech materials, as well as familiarity with library science. During the next three years, the Library employed scores of persons with backgrounds in science and engineering. The sci-tech reports introduced the need for a change in cataloging procedures. Also, demands for copies of those reports, abstract and index bulletins, and catalog cards stimulated the Library to expand its reproduction facilities and search for new duplication techniques.

In 1949, the Library of Congress established the Science Division and the Air Information Division to do special cataloging, abstracting and other bibliographic tasks largely in science and technology for the Department of Defense. In 1950 it developed a bibliographic project, SIPRE, for the Snow, Ice, and Permafrost Establishment of the U. S. Army Corps of Engineers. In 1952, the Technical Information Division was formed by combining the Navy Research Section, SIPRE, and some other bibliographic activities supported by federal agencies other than the Library.

Although the time tables of changes were not the same and the new special services were different, federal libraries such as the Army Medical Library and the libraries of the Departments of Agriculture and Interior experienced analogous growths in collections and responded similarly to new demands by inaugurating special services that required technical innovations. For example:

1. In 1942, the Army Medical Library incorporated photoduplication services into its operations; in 1943 it decided to supply microfilm copies without cost to government agencies and authorized investigators; and in 1950 it funded a Johns Hopkins research project to study problems of medical bibliography with emphasis on investigating application of machine methods.

2. In 1942, the Department of Agriculture Library started publication of the *Bibliography of Agriculture;* in 1946 it began to supply copies of articles cited in *Chemical Abstracts* to members of the American Chemical Society and subscribers to *Chemical Abstracts*; and in 1952 it introduced the rapid selector into its library operations.

During the same period in which the major federal libraries were experiencing growth in collections and requests for new services, federal R&D agencies developed an entirely new type of information service—the technical information center. The major stimulus that led to the organization of these centers was the need for rapid and widespread dissemination of R&D results. Other considerations included the dispersion of R&D centers and the need to acquire and exploit results of foreign research efforts. Most major technical information centers were created by R&D agencies born during or just after World War II. Earlier mention was made of the Library of Congress's Science and Technology Project. This activity was proposed to the Library of Congress and funded by the Office of Naval Research, which was established in 1946. Another technical information service, the Air Documents Division, which became the Central Air Documents Office and later a part of the Armed Services Technical Information Agency was initiated by research and development components of the Air Force, which itself was created as a separate federal agency during the war. ASTIA was established in 1959 by the Department of Defense, which was then only a few years old. As recounted earlier, ASTIA became the Defense Documentation Center in 1965 and soon thereafter was placed under the Defense Supply Agency, another federal agency created after World War II.

Three other large federal technical information services came into being after World War II. One was an outgrowth of the sci-tech information activities of the Manhattan District Project, which, as recounted earlier, became the Technical Information Section of the U. S. Atomic Energy Commission after it was established in 1946. The second technical information service was the Office of Technical Services. The major stimulus for the Department of Commerce to establish OTS came from the Publication Board, an interdepartmental committee created in 1945 to facilitate the declassification of foreign and domestic sci-tech documents and to encourage their wide distribution. The third and youngest of the large technical information services established during the 1942-1972 period was the Office of Scientific and Technical Information of the National Aeronautics and Space Administration; it was established in 1960. NASA itself had been created only two years before as the successor agency to the National Advisory Committee for Aeronautics.

Although a major purpose of all these information services was to speed up and improve the distribution of sci-tech reports, either initially or later they were given additional tasks. All placed high priority on gaining quick access to reports of interest to their sponsoring agencies. Because of the need to distribute copies of documents to widely dispersed users, these services developed mechanisms for rapid reproduction and distribution of documents; all also produced such bibliographic tools as catalog cards, abstracts, indexes, and special bibliographies. An added burden was maintaining strict control over access to documents that were security classified.

These centers were constantly in search of new technical devices and methods. They revised traditional cataloging procedures to make them more responsive to users of their products. Experiments were conducted with different indexing and subject classifying methods. Although these technical information services

Trends, Persons, and Future Directions   151

were not the first to employ microreproductions of documents, they eagerly explored ways to make these forms more useful. For example, the Central Air Documents Office and its successor, the Armed Services Technical Information Agency, experimented with the use of microfilm in reels, in small capsules, and in sheets, and the AEC Technical Information Section distributed documents in microcard form. Most agencies used new reproduction techniques and quickly adopted the electrostatic process when it became available in 1956. Techniques for using punchcards in automatic data processing and photo-offset reproduction were refined through experimentation and use by these services. All were also interested in improving user knowledge and use of their products. For example, CADO and the Library of Congress's science and technology projects made use of liaison officers to visit R&D establishments to explain their services and products and AEC/TIS had its Conference of Librarians. All of these agencies welcomed opportunities to describe their services and products in meetings with librarians, information specialists, and scientists and engineers. In addition to making its services and products available to R&D laboratories, each service established a network of regional centers that were furnished with its documents, indexes, and other bibliographical tools. CADO, and later ASTIA, also organized branch outlets in metropolitan areas where use of its products was extensive. From time to time, all services issued brochures describing their products and services and explaining how to procure and use them.

By the late 1950s, and particularly in the early 1960s, these centers as well as many other libraries and information centers began to explore possible ways to use computers and telecommunications. Microfiche and electrostatic reproduction had already been adopted for their programs. The first computer-aided operations were directed to storing and organizing index entries or similar files. Later, full bibliographic entries and abstracts were included. Concurrently, the services were engaged in systems planning and in design and development that included use of computers, telecommunications, rapid copying techniques, and microfiche. To begin with, emphasis was on selective-dissemination services, back-stopped by computer-based operations that could organize and produce indexes, bibliographies, catalogs, and other special tools including distribution lists, thesauri, and current inventories of research projects.

The following examples are illustrations of trends toward use of new techniques (no attempt is made to identify the first use of new equipment nor to name all agencies using similar techniques or systems):

1945-1948: Library of Congress, OTS, Department of Agriculture, and military information agencies were making extensive use of microfilm for storage and distribution of documents.

1948: Patent Office was developing a coordinate indexing system and beginning to use processing equipment employing punchcards for search of selected classes of patents.

1949: AEC/TIS used an IBM punchcard system for preparing indexes and special bibliographies.

1952: Department of Agriculture was using the rapid selector and photo-clerk in its acquisition and cataloging activities.

1952: AEC/TIS used microcards for storage and distribution.

1955: Patent Office and National Bureau of Standards experimented with the Bureau's SEAC computer to develop search strategies for identifying patents.

1956: ASTIA's Wright Air Force Base Center was using seven electrostatic (Xerox Copyflow) reproduction units to produce copies of reports.

1959: AEC was preparing current and cumulative indexes to *Nuclear Science Abstracts* using IBM punchcards and Listomatic camera.

1959: ASTIA used a UNIVAC SS-90 for some of its bibliographic processing.

1960: AEC entered bibliographic data on magnetic tape in preparing indexes and bibliographies.

1960: National Library of Medicine used punchcards and the step-camera to produce *Index Medicus.*

1962: Documentation Incorporated began operating a computer-based information processing center for NASA.

1962: NASA used 5"x8" microfiche in its user-service operations.

1962: AEC stored its bibliographic data on magnetic tape and initiated selective dissemination service (RESPONSA).

1963: Remington-Rand installed a Univac 1107 bibliographic system for ASTIA.

1963: AEC replaced microcard with microfiche.

1963: NASA first offered a selective dissemination service (SCAN).

1964: NLM used MEDLARS system with GRACE component to produce *Index Medicus.*

1965: NLM initiated its recurring bibliographies and demand searching services.

1965: ASTIA began having its *Technical Abstract Building* (TAB) printed at the Government Printing Office using Linofilm composer.

1966: NASA accepted DDC bibliographic data on magnetic tape.

1966: Library of Congress's MARC (machine readable catalog) program began distribution of tapes.

1967: Clearinghouse for Federal Scientific and Technical Information began introducing bibliographic data into DDC's computer system from remote consoles.

1968: Library of Congress's MARC II began operation.

1969: AEC began to enter subject headings and index entries into its computer from consoles.

1969: NASA was using an on-line interactive computer system (RECON).

1969: National Agricultural Library initiated sale of CAIN computer tapes.

1971: National Library of Medicine furnished medical information to hospitals and other units in Alaska via satellite communication network.

The above list of technical innovations illustrates the following trends in federal agencies' use of new technologies: Federal libraries and information agencies were using punchcards and microfilm in their preparation of catalogs, indexes, and bibliographies during the late 1940s and early 1950s. Not until the late 1950s, however, had these techniques been sufficiently refined for use operationally in the preparation of abstract and index publications with current and cumulative indexes. By 1960 some federal information agencies were using magnetic storage devices for bibliographic data and employing computers and auxiliary equipment to prepare indexes and bibliographies. In the middle of the 1960 decade, a few agencies were using computer-controlled composing equipment with multiple type fonts. At that time several libraries and information centers were able to offer selective-dissemination bibliographic services, but not until the end of the decade were on-line-interactive, computer-based, bibliographic information systems in operation.

A trend that is not shown clearly in the above listing, is the extent to which one agency could make use of another's data recorded in magnetic form. NASA was accepting DDC's data in this form in 1966. The Library of Congress's MARC system was predicated on the use of its tapes by other organizations for their own systems. Interchange of these types of bibliographic data stored in the magnetic format continued to increase. However, problems related to formatting materials, differences in data elements, methods of indexing, and so forth required "translation" of these magnetically recorded data. Problems of this type have impeded widespread exchange of such data.

Strongly influencing the federal library and information services move toward automation was the rate of development of the software and hardware necessary for information processing. No attempt is made here to review this important aspect of information activities. It should be noted, however, that information services adapted these intellectual and mechanical devices to their use soon after they became readily available and also made extensive use of microforms and rapid reproduction devices as soon as they appeared. Many agencies were responsible for many of the refinements and innovations in this type of equipment and the systems that were developed to use them.

The federal information services made little use of the first generation of computers with their limited memories, lack of auxillary equipment, and relatively slow speed. The second generation of computers were adapted to aid in the storage and manipulation of bibliographic data for preparation of catalogs, abstracts, and bibliographies. When the third generation of computers became available, the information services began to develop innovative user services such as selective dissemination and on-line remote use of centralized files. Improvement in performance and reduction in unit cost of telecommunications materially aided the information community to develop systems for remote access to information files. Satellite and microwave developments have and will continue to allow federal information agencies to further extend services to remotely located organizations and individuals.

Two pressures are increasingly affecting sci-tech information services. One is the need for standards that will allow easy exchange of information between different data bases. This standardization must be in two areas: first, the physical arrangement of materials and, second, the intellectual organization of indexes and subject headings. Although files do not have to be identically organized and in the same format, they must be compatible and consistent enough to allow for easy translation. The second pressure comes from the increasing quantity of available information and the expanding complexity of its uses. The impact of these phenomena require information agencies to develop cooperative projects for collecting and organizing information, as well as services to meet a wide variety of user requirements.

The above discussion leads to another facet of the information industry, namely, the need for better understanding of how information should be managed. More reliable data is needed on: unit cost of operations; identification of user requirements; methods for evaluating the efficiency and effectiveness of information systems; and programs to inform users about available information sources and services and how to make the best use of them. Some work has been done in each of these areas but much remains to be done.

FEDERAL PERSONNEL INFLUENTIAL IN SCI-TECH INFORMATION

In an organization as large as the federal government, scores of persons have influenced both the direction and quality of its sci-tech information programs, and thousands more have been involved in operating the services. Federal sci-tech information services are what they are today because of these persons. Directors of information programs have had to persuade policymakers and administrators of the necessity and worth of these activities. The directors have had to prepare and defend the budgets for their programs in a climate where many different federal programs were competing vigorously for available federal dollars, and they have had responsibility for the effectiveness of the information activities.

During the 1942-1972 period, many persons contributed to the advancement of federal sci-tech information programs. Within this group are a smaller number who participated as administrators of two or more sci-tech information services, for there was considerable mobility among federal information services personnel. The selection of the persons listed in this section is arbitrary and based largely on the criteria that each one participated in the development of a new federal program or in a marked expansion or change in an on-going service. Since there are few published or unpublished records of individual contributions to the development of federal sci-tech information services, this author has had to draw upon his own experience and the suggestions from other persons who had long and varied careers in the information services field. The brief resumés of these persons indicate that federal information administrators came to this field with a wide variety of experience and training; that the requirements for successful performance in different administrative posts were sufficiently similar so that a person could transfer from one information program to another and continue to perform effectively; and that mobility of administrators has resulted in extensive and valuable exchange of experience among the federal sci-tech programs. Although many persons outside the federal establishment contributed to the advancement of federal sci-tech information programs, they have been omitted from this listing of federal information services administrators.

Scott Adams held important administrative positions in the National Library of Medicine or its predecessor agencies. During his last nine years in the federal government, he was deputy director of the National Library of Medicine. He transferred to the National Science Foundation in 1960 as program director for foreign science. In this position, he was instrumental in initiating the PL-480 program of translations with the use of U. S. surplus foreign credits. In 1961, he returned to NLM, and in 1969 he retired from federal service to be technical advisor to the UNISIST Central Committee. As deputy directory of NLM, he was involved in the planning and development of the MEDLARS program.

Burton W. Adkinson was chief of the Maps Division of the Library of Congress in 1948 when he was appointed assistant director of the Reference Department. In 1949 he became its director. He was directly involved in the establishment of the Navy Research Section, the SIPRE project, the Air Information Division, the Science Division, and the Technical Information Division, and the combination of the last two to form the Science and Technology Division. In 1957, he transferred to the National Science Foundation as head of the Office of Scientific Information. When the Office of Science Information Service was established in 1958, he was its first director. In 1971, he retired from this position to become the director of the American Geographical Society.

Andrew A. Aines, prior to 1962, was director of Army Technical Information in the Office, chief of R&D, Army Research Office. In late 1962, he transferred to the Defense Directorate of Research and Engineering and became executive director of the Committee on Scientific and Technical Information. He transferred to the Office of Science and Technology in 1964 as a technical assistant. In 1966 he became chairman of COSATI, while continuing to perform the technical assistant's duties in sci-tech information. He has been the U. S. representative to the OECD Information Policy Group since 1967. In 1971, he transferred to NSF's Office of Science Information Service, as special assistant to the head of the office. In that year he also was appointed a member of the National Commission of Library and Information Science.

Samuel Alexander was employed in the Bureau of Standards at the end of World War II. He was involved in the development of the early models of computers. As part of this assignment, he worked with the Bureau of Census in its development of a computer system to handle census data. About 1949, he was placed in charge of the Bureau's computer services. During all of the above period, Alexander headed a group developing the computer know as SEAC. He and his staff began in 1955 a cooperative program, known as HAYSTAQ, with the Patent Office. This project was to develop auxillary equipment and strategies for search of patent files. As chief of the Bureau's Computer Division, in 1959 he established a Research Information and Advisory Service for Information Processing. During all the above period Alexander acted as a formal or informal information processing consultant to several federal agencies, including the Census Bureau, the Patent Office, the National Science Foundation, the BioSciences Information Exchange, and a number of defense and intelligence agencies.

Edward L. Brady transferred from private industry to the National Bureau of Standards in 1963, where he had responsibility for developing the National Standard Reference Data System. Recognition of his effectiveness in this assignment led to his being named in 1969 associate director for Information Programs of the National Bureau of Standards.

Edward M. Brunenkant became the first chief of the Industrial Information Section of AEC's Technical Information Service in 1957. In 1960, he was appointed chief of AEC's Technical Information Division. Brunenkant headed the interdepartmental group that developed the microfilm and microfiche standards that were adopted by the Federal Council for Science and Technology. As chief of TID, he encouraged the development of the International Nuclear Information Service of the International Atomic Energy Agency, and began adapting the NASA's RECON system to AEC's information activities.

Lee G. Burchinal was brought to the government by the Office of education in 1965 to establish an information program in support of research in education. The Education Resources Information Centers (the ERIC system) is the result of his planning and efforts. As part of the ERIC program he provided leadership in developing a low-cost, portable microfiche reader and stimulated the establishment of state-wide services to use the ERIC resources and services.

Walter M. Carlson came in 1962 to the Office of the Director of Defense Research and Engineering. Previously he had been involved in technical information activities in the duPont Corporation. Under Carlson's direction, the Office of the Director of Technical Information was established in the Department of Defense. He was also involved in the change of ASTIA to the Defense Documentation Center and encouraged the Department of Defense to clarify the methods of support and the roles of the Information Analysis Centers which the department supported. In 1966, he left the Department of Defense for a position with IBM.

Verner W. Clapp, as director of the Processing Department and chief assistant librarian of Congress, was involved in the development of the Library's photoduplication services and the use of microfilm for storage and service. He also participated in the development of Library's sci-tech projects funded by other federal agencies. In 1955, he left the Library of Congress to become the first president of the Ford Foundation's Council of Library Resources. From this position, he continued to participate actively in advancing developments in libraries and sci-tech centers both within and outside the federal government.

Ruth E. Davis has been involved in federal activities in computer and information sciences and technology since 1961 when she was staff assistant in the Office of the Director of Defense Research and Engineering. In 1967 she became associate director for R&D in the National Library of Medicine. In addition, in 1968 she became the first director of the Lister Hill National Center for Biomedical Communications. In 1970, Davis transferred to the National Bureau of Standards as the director of the Center for Computer Sciences and Technology. Under her leadership, this center was raised to an institute of the Bureau in 1972.

Melvin S. Day was a chemist in the Manhattan District Project in 1946 when he joined the AEC technical information program at Oak Ridge. From 1946 to 1955, he held several positions in the AEC Oak Ridge extension. In 1956 he became chief of the extentions activities, and in 1958 he moved to AEC's Washington, D. C., headquarters to direct the Technical Information Division. In 1960, Day transferred to NASA as deputy director of the Office of Technical Information and Education. When NASA established a separate Technical Information Division in 1962, he became its first director, and in 1966 he was appointed deputy assistant administrator for Technology Utilization. Day became Head of NSF's Office of Science Information Service in 1971 and then transferred in 1972 to the National Library of Medicine where he continues as deputy director.

Stella L. Deignan was director of the Office of Exchange of Information, National Institutes of Health, in 1950 when she transferred to the National Research Council and became the first director of the Medical Sciences Information Exchange. This inventory of current research projects in the medical sciences became so useful under her leadership that in 1953 its scope was enlarged to cover all the biosciences, and some branches of psychology. In 1959, when it was apparent that the Biological Sciences Information Exchange's scope would be enlarged to cover all fields of science, Deignan resigned to take a position with the World Health Organization.

Bernard M. Fry was in the Manhattan District Project as head of the library unit at Oak Ridge National Laboratory. In 1946, when AEC was established, this library operated a very successful declassification program for AEC. In 1955, he became director of AEC's Technical Information Service. In 1959 he transferred to the Office of Science Information Service in NSF. As deputy head of this office, he was influential in developing its program. In 1964, he was asked to do a study of the Office of Technical Service for the Department of Commerce. A year later he became director of the Office of Technical Services, which a short time later became the Clearinghouse for Federal Scientific and Technical Information. In 1967, he left federal service to become professor of library and information sciences at Indiana University and dean of its School of Library Science.

Dwight E. Gray contributed to several sci-tech programs in the government. He came to AEC in early 1950 where he worked in an editorial and writing capacity. In late 1950, he became chief of the Navy Research Section in the Library of Congress. When the Library's Technical Information Division was established in 1952, he was its first chief. In 1955, Gray transferred to the Office of Scientific Information in NSF, where he developed the Government Research Information Program. He held several responsible positions in the Office of Science Information Service. In 1963 he returned to the Library as chief of the Division of Science and Technology. In 1965 he retired from the Library to become Washington representative of the American Institute of Physics.

John C. Green held several positions in the Department of Commerce that influenced federal sci-tech programs. When the Publication Board was established in 1945, he was executive secretary of both the Inventor's Council and the Interdepartmental Committee for Reclassification of Scientific Information. He continued to hold these positions after he was appointed the first director of the Office of Declassification and Technical Services, which later became the Office of Technical Services. He held this position until 1963. In addition, the Department of Commerce placed Green on several departmental committees investigating sci-tech information programs. One of these was a group set up to advise the Secretary of Commerce on steps to implement the Kelly and King committees' recommendations on the reorganization and improvement of the Patent Office's information system.

William T. Knox became technical assistant to the director of the Office of Science and Technology in 1964. Much of his effort was directed toward improving coordination and cooperation among federal information agencies and in participating in international information activities. Soon after he took office, he became chairman of COSATI and the National System Task Force of that committee. He also prepared the basic paper that led to the establishment of the Organization for Economic Cooperation and Development's Information Policy Group. In 1966 he left federal service, but returned in 1971 as director of the newly established National Technical Information Service, which he continues to direct.

W. Kenneth Lowry was in the Office of Technical Services before he joined the Library of Congress in 1948 where he was administrative officer in the Science and Technology Project. In 1949 he headed the project for a while, then transferred to the Department of the Army as director of its libraries. About 1953 he assumed responsibility for technical information in the Air Research and Development Command in Baltimore. In this position, he had considerable influence on the policy and programs of the newly formed Armed Services Technical Information Agency. He left this position in 1956 to become director of libraries of Bell Telephone Research Laboratories. Since then, he has served on several federal advisory committees.

Eugene Miller was recruited in 1945 by the National Advisory Committee for Aeronautics to initiate a scientific and technical report program and to develop coordination among the NACA information services located in its laboratories. In 1951, he became deputy directory of the newly formed Armed Services Technical Information Agency; in 1952, he left this position to become vice president of Documentation Incorporated.

Foster E. Mohrhardt entered federal service in 1946 as chief of the Library and Reports Division of the Office of Technical Services. He left federal service in 1947 but continued as a contract consultant to Brookhaven National Laboratory. In 1948 he became chief of the Library Division of the Veterans' Administration. In 1954, he was appointed director of libraries in the Department of Agriculture. During his administration, it was raised to the status of a national library. Also, during his regime, the basic planning and design for automated information services for agriculture were initiated. In 1968 he retired from the federal government to take a position with the Council on Library Resources.

Frank B. Rogers was assigned the directorship of the Army Medical Library in 1949, at which time he was a colonel in the Army Medical Service. During Rogers' term of office, which extended to 1963, the library became, in 1952, the Armed Services Medical Library, and in 1956, the National Library of Medicine. During this period, many changes were made in the products and services, such as providing microfilm in lieu of loan of materials whenever feasible and mechanizing the production of the Monthly *Index Medicus,* which eliminated the need for *Current List of Medical Literature* and the *Quarterly Index Medicus.* Rogers was heavily involved in the design and development of MEDLARS, which began operations in 1964, the year that Rogers became director of the Medical Library of the University of Denver Medical School.

Hubert Sauter joined NASA's Office of Scientific and Technical Information in 1961, where he participated in developing that office's program. In 1966, he transferred to the Clearinghouse for Federal Scientific and Technical Information as deputy director and became its director in 1967. In 1972, he was appointed assistant director for Program Direction and Evaluation of the National Technical Information Service. That same year he moved to the Defense Documentation Center and became its administrator in July. He continues to direct that agency's operations.

Ralph R. Shaw, in 1940, became director of libraries in the Department of Agriculture and was on military leave in 1944 and 1945. He developed the photocharger, the photoclerk, and the rapid selector for use in the department's libraries. He planned and started the *Bibliography of Agriculture.* Before and after his military service, he participated in several interdepartmental committees that planned new federal library and other information services, one of which was the Office of Technical Services in the Department of Commerce. He left federal service in 1954 to become professor of library science at Rutgers University.

John Sherrod was recruited by the Library of Congress in 1950 to head the SIPRE project. In 1954, he became chief of the Science Division. In 1963, he transferred to the Technical Information Division of AEC as deputy director. After five years with this division, he became director of libraries in the Department of Agriculture. In 1973, Sherrod left the federal government to become director of NASA's Information Processing Center, which is operated by Informatics, Inc.

John F. Stearns served in responsible positions in four different federal sci-tech agencies. He came to the Library of Congress in 1949 as assistant chief of the Air Information Division when it was formed and soon became its chief. Later he served as chief of the Aeronautics Section. About 1954, he transferred to the Technical Information Services of the Air Force Research and Development Command. In 1962 he became the first director of the Library of Congress National Referral Center for Science and Technology, after having been in ASTIA and NASA's Office of Scientific and Technical Information. Later he returned to NASA's technical information program and then in 1971 transferred to NSF's Office of Science Information Service from which he retired in 1973.

Robert B. Stegmaier, Jr., was a staff officer in the Office of the Assistant Secretary of Defense for Research and Development for many years. During this time, he had considerable responsibility for the planning and establishing of the Armed Services Technical Information Agency in 1951. From then until 1963 his staff positions carried responsibility for developing ASTIA's policies. In 1963, he became administrator of the Defense Documentation Center, where he developed its program and directed its operations until 1974, when he retired.

Mortimer Taube was assistant director of the Processing Department of the Library of Congress when he was appointed head of the Science and Technology Project in 1947. He organized this new project's program and staff. In 1949, he transferred to AEC as deputy director of the Technical Information Section. He left AEC about 1952 to form his own company, Documentation Incorporated.

Alberto Thompson became involved in sci-tech information activities during the Manhattan District Project. In 1946 he was appointed first chief of AEC's Technical Information Section. In 1955, he transferred to the National Science Foundation as head of its Office of Scientific Information—a position he held until his death in 1957.

As stated before, this list is a highly selective and arbitrary one of persons who have substantially influenced the federal government's sci-tech programs and policies. Numerous other persons who have made significant contributions to the development of information services have been mentioned throughout this text in relation to specific plans, programs, and policies. Many others have not been mentioned because their contributions, while important, were somewhat less integral to the evolution of federal policies and programs. The members of the foregoing list were instrumental in the development of federal sci-tech information services from largely manual operations to on-line, interactive systems.

POSSIBLE FUTURE DEVELOPMENTS

This final part of the discussion concentrates on future trends in federal programs and policies. No attempt is made to forecast specific future technical advances even though these innovations do influence federal sci-tech programs and policies.

One trend in federal sci-tech programs during the past decade, which will continue, is the move toward consolidation and coordination (see Appendix G for the conclusion of a Senate special subcommittee report issued in 1975). A significant factor in this trend has been the federal government's actions to clarify and consolidate federal R&D programs. For example, during the past ten years the Department of Commerce has developed large coordinated R&D programs in atmospheric and oceanographic sciences. As part of this consolidation and coordination process, R&D units from other federal agencies have been transferred to the National Oceanographic and Atmospheric Agency in the Department of Commerce. This agency has moved to develop a coordinated sci-tech information program to service this large and varied R&D effort. It is called the Oceanic and Atmospheric Scientific Information System. This system includes the information activities of the Weather Bureau, the Coast and Geodetic Survey, the National Marine Fisheries Service, and several that were transferred—the National Oceanographic Data Center for the Navy Department, and services in support of the Sea Grant Program from the National Science Foundation.

With the establishment in 1975 of the Energy Resources Development Agency, the former technical information activities of AEC are being expanded to include information on all sources of energy. This move is being accomplished by having ERDA's information program solicit cooperation and attempt to effect coordination with both private and federal information services that have information and data on energy.

The 94th Congress had indicated additional interest in improved administration and coordination of federal R&D, including sci-tech information. Both Senate and House Committees have conducted studies and issued reports. Also the following bills have been submitted to the House for consideration; HR 4461, 9058, and 10230. The latter two bills include the following statement: "It is a responsibility of the federal government to insure prompt, effective, reliable, and systematic transfer of science and technology information by such appropriate methods as funding technical evaluation centers, cost sharing of information dissemination progress conducted by such nongovernmental organizations as industrial groups and technical societies, and assistance in publication of properly certified science and technology information." These trends of administrative or legislative actions to consolidate or coordinate federal sci-tech information activities will continue.

Concurrent with administrative consolidation have been moves toward coordination and cooperation among federal information agencies. To illustrate this trend: The three national libraries—the Library of Congress, the National Library of Medicine, and the National Agricultural Library—are developing a joint serials data program; agencies are increasingly using each other's bibliographic and numerical data bases; and AEC (now ERDA) has adapted NASA's RECON system for its own use. As information technology improves, this coordination and cooperation will be increasingly important among federal information services.

The federal government recognizes that improved coordination is facilitated by the development of standards. In order to further and accelerate development along this line, the federal government has assigned the Bureau of Standards the responsibility for establishing federal standards for information processing. The Bureau has delegated the task to its institute for Computer Sciences and Technology. Among the institute's program objectives are the following: provide scientific and technological advisory services to federal agencies, recommend federal information processing standards, and participate in the development of voluntary automatic data processing (ADP) standards.

In another move to accelerate the development of standards in the library and information sciences, the Office of Science Information Service of the National Science Foundation partially supports the Z39 Committee of the American National Standards Institute, which is responsible for the formulation of standards. In the future, standardization of basic elements of information processing of all types will become increasingly important as national systems and networks evolve.

This leads to another area where developments are rapidly taking place. Earlier discussions showed how federal agencies are moving toward networks of information systems. The National Library of Medicine, the Library of Congress, NASA, the National Agricultural Library, and DDC are all working toward on-line interactive systems that can be accessed through consoles. Improvement and expansion of telecommunications and satellite transmission will further this movement.

The present computerized information systems have concentrated on use of bibliographic and numerical data. As computer and communications technologies improve in efficiency and economy, the use of textual and graphic materials in these systems will expand. Editing and organization of textual material with the aid of the computer and console is increasing. Architectural designing and other modelling is being done with the aid of computers and auxillary equipment. Publishing is depending increasingly on computer-controlled composition. All these developments will force a marked change in the role of the scientific journal and monograph during the next decade. Heavy reliance on printed media for dissemination of information is decreasing as techniques for the use of computers and electronic communication improve. The decade of the 1980s should see a marked change in the role of the scientific journal and monograph.

Federal agencies have cooperated with private organizations who have developed computerized systems that allow access to multiple files. Such organizations as the Lockheed Corporation and Systems Development Corporation have become what might be termed "information brokers." Through leasing and other arrangements with federal and nonfederal organizations, these corporations have assembled many large files that can be addressed via consoles located throughout the United States and have devised methods for allowing customers to be aware of the per unit cost before using the systems. Federal information agencies are exploring this method of information handling, and the number of "information brokers" will undoubtedly increase during the coming years.

In summary, federal agencies over the next decade will increasingly consolidate and coordinate their information systems. Programs will be developed to take advantage of the constantly improving computer, microform, and communications technologies. R&D and other federal personnel will have ready access to a wide variety of information files through consoles located throughout the United States. National network of major federal agencies will allow for easy use of these information resources by individuals and organizations both in the United States and abroad.

Bibliography

Adams, Scott. *Information for Science and Technology: The International Scene.* Occasional Papers, No. 109. University of Illinois, Graduate School of Library Science. Urbana, November 1973, pp. 45 multilith. ISSN 073-5310.

American Institute of Physics, Information Division. *A Program for a National Information System for Physics and Astronomy 1971-1975.* American Institute of Physics, New York, June 1970, pp. 60, plus 4 appendixes, pp. 37.

Bolt Beranek and Newman, Inc. *Toward the Library of the 21st Century.* A Report on Progress made in a Program of Research sponsored by the Council on Library Resources. Bolt, Beranek and Newman, Inc., Cambridge, Mass., 1964, pp. 30.

Council on Library Resources, Inc. *10th Annual Report for Year Ended June 30, 1966.* Council on Library Resources, Inc., Washington, D. C. 1966, pp. 128. Pages 9-31 contain a summary by Verner W. Clapp of CIR activities from 1956.

Clawson, Marion. *The Bureau of Land Management.* New York: Praeger Publishers, 1971, pp. 209.

Cochrane, Rexmond, C. *Measures for Progress: A History of the National Bureau of Standards.* U. S. Department of Commerce, National Bureau of Standards, Washington D. C., 1966 (second printing 1974), pp. 703. Printed by U. S. Government Printing Office.

Conover, Milton. *The General Land Office: Its History, Activities and Organization.* Series Monograph No. 13, Institute for Government Research. Baltimore: The Johns Hopkins University Press, 1923, pp. 224.

Chemical Abstracts Service. *CAS Today: 60th Anniversary Edition.* Chemical Abstracts, Columbus, O., 1967, pp. 32. Facts and data about the Chemical Abstracts Service.

Gray, Dwight E. "Science Division—Library of Congress." *Physics Today* 4 (October 1950): 417-24.

_____. "Scientific Liaison Offices." *Physics Today* 4, no. 3 (February 1951): 28-29.

_____. "An Experiment in Standardization." *Physics Today* 4, no. 3 (March 1951): 8-9.

_____. "Recent Developments in Physics Abstracting." *Physics Today* 4, no. 8 (August 1951): 18-20.

_____. "Dissemination of Technical Information by AEC." *Physics Today* 4, (November 1951).

_____. "Office of Technical Services of the Department of Commerce." *Physics Today* 4, no. 12. (December 1951): 24-26.

_____. Scientists and Government Research Information: College and Research Libraries. *Physics Today* (January 1957): 23-27.

_____. "Information and Research-Blood Relatives or In-Laws?" *Science* 137 (July 27, 1962): 263-66.

Gray, Dwight E., and Johnson, J. Burlin. "Services to Industries By Libraries of Federal Government Agencies." *Library Trends* 14, no. 3 (January 1966): 3332-46.

Holt, W. Stull. *The Bureau of the Census: Its History, Activities and Organization.* Service Monographs of the U. S. Government No. 53, Institute for Government Research. Washington, D. C.: The Brookings Institution, 1929.

Japan Documentation Society (NIPDOK). *Science Information in Japan.* 2nd Rev. Ed., Haruo Ootuka, Chairman of Editorial Committee. Japan Documentation Society, Tokyo, 1967, pp. 192. Printed by Kikai Sinko Kaiken, Siba Park, Tokyo.

Japan Information Center for Science and Technology. *The Japan Information Center for Science and Technology.* The Japan Information Center for Science and Technology, Tokyo, 1967, pp. 10. Contains typed inserts of updated material to 1970.

The Johns Hopkins, University. "The Information Deluge." Special issue of *The Johns Hopkins Magazine* (Fall 1967, pp. 33. Includes articles by J. C. R. Licklider, William D. Garvey and Bertita E. Compton, Ferdinand F. Leimkuhler and Anthony E. Neville, Martin Greenberger and William E. Passano.

King, Gilbert, et al. *Automation and the Library of Congress.* Report of the Committee to the Librarian of Congress; survey sponsored by the Council on Library Resources, Inc. Library of Congress, Washington, D. C., 1963, pp. 88.

National Academy of Science/National Research Council, Information Systems Panel, Computer Science and Engineering Board. *Libraries and Information Technology: A National System Challenge.* A Report to the Council of Library Resources, Inc. National Academy of Science/National Research Council, Washington, D. C., 1972.

National Research Council of Canada, National Science Library of Canada, *Annual Report 1972-1973.* National Science Library, Ottawa, 1973, pp. 36. In both English and French.

Unesco, Division of Scinetific and Technical Documentation. *UNISIST— Newsletter.* Unesco, Paris, for 1972 to 1975. ISSN 0300-2519.

U. S. Atomic Energy Commission. *Directory of USAEC Information Analysis Centers.* AEC Technical Information Center, Oak Ridge, Tenn., October 1974, pp. 35.

_____. *AEC Technical Information Center—Its Functions and Services.* AEC

Technical Information Center, Oak Ridge, Tenn., October 1972, pp. 21.
_____. Technical Information Program. *A Bibliography*. AEC Technical Information Center, Oak Ridge, Tenn., May 1972, pp. 18. TID-3326.
U. S. Dept. of Agriculture, National Agricultural Library. *A Progress Report on the Agricultural Science Information Network*. National Agricultural Library, Beltsville, Md., 1971, pp. 14, multilith.
U. S. Congress, House of Representatives, Select Committee on Committees. *The Congress and Information Technology*. Staff Report of Congressional Research Service. Library of Congress, Washington, D. C., May 5, 1974.
U. S. Department of Commerce, National Bureau of Standards. *National Bureau of Standards.* National Bur. Stand. (U. S.) Special Pub. 397. National Bureau of Standards, Washington, D. C., 1974, pp. 64. Printed by the U. S. Government Printing Office.
_____. *Federal Information Processing Standards, Index*. National Bur. Stand. (US) Fed. Inf. Process. Stand. Publ. (FIPS Pub) 12-2. National Bureau of Standards Washington, D. C. 1974, pp. 195.
U. S. National Aeronautics and Space Administration. *Semiannual Report to Congress*. NASA, Washington, D. C., various years. Printed by the Government Printing Office. Each report has a section on information programs.
U. S. National Science Foundation. *Nonconvention Scientific and Technical Information Systems in Current Use*. National Science Foundation, Washington, D. C., 1958-1966. There were four numbered issues of this publication; the indexes in each volume identify performers, equipment, and organizations.
_____. *Science Information Notes*. National Science Foundation, Washington, D. C., 1959-1968. Printed by the Government Printing Office. This bimonthly publication gives brief notes on activities in the library and science information field.
_____. "Making Technical Information More Useful: The Management of a Vital National Resource." Unpublished report known as "Greenberg Report." National Science Foundation, Washington, D. C., June 1972.
_____. Scientific Information Activities of Federal Agencies Series. National Science Foundation, Washington, D. C., 1958-1966. Printed by the Government Printing Office. Each pamphlet (listed below) was reviewed by the agency discussed for accuracy of the information.

No. 1, Department of Agriculture (October 1958) NSF 58-27.
No. 2, Office of Naval Research (May 1959) NSF 59-19.
No. 3, Department of Commerce, pt. 1 (November 1959) NSF 59-58.
No. 4, U. S. Government Printing Office (March 1960) NSF 60-9.
No. 5, Tennessee Valley Authority (August 1960) NSF 60-44.
No. 6, National Science Foundation (October 1960) NSF 60-56.
No. 7, Department of Commerce, pt. 2 (October 1960) NSF 60-58.
No. 8, Department of Commerce, pt. 3, National Bureau of Standards

168 Bibliography

(October 1960) NSF 60-59.
No. 9, Federal Communications Commission (May 1961) NSF 61-12.
No. 10, Veterans Administration (June 1961) NSF 61-22.
No. 11, Treasury Department (November 1961) NSF 61-8.
No. 12, Department of Interior, pt. 1, Bureau of Reclamation, Office of Saline Water, Fish and Wildlife Service, National Park Service, Bonnevile Power Administration, Bureau of Indian Affairs, Bureau of Land Management (December 1961) NSF 61-77.
No. 13, Smithsonian Institution (June 1962) NSF 62-8.
No. 14, Federal Aviation Agency (August 1962) NSF 62-19.
No. 15, U. S. Air Force, pt. 2, Office of Aerospace Research (September 1962) NSF 62-32.
No. 16. Department of Interior, pt. 2 (October 1962) NSF 62-35.
No. 17, U. S. Air Force, pt. 4, Air Force Systems Command (May 1963) NSF 63-16.
No. 18, U. S. Atomic Energy Commission (October 1963) NSF 63-38.
No. 19, U. S. Navy, pt. 1 (November 1963) NSF 63-42.
No. 20, U. S. Navy, pt. 2, Office of Naval Research (November 1963) NSF 63-43.
No. 21, Department of Health, Education and Welfare, pt. 1, (December 1963) NSF 63-50.
No. 22, U. S. Air Force, pt. 1 (January 1964) NSF 63-45.
No. 23, Department of Health, Education and Welfare, pt. 2, Public Health Service (February 1964) NSF 63-46.
No. 24, Department of Health, Education and Welfare, pt. 3, National Institutes of Health (February 1964) NSF 63-52.
No. 25, U. S. Navy, pt. 3, Bureau of Ships, Bureau of Naval Weapons (February 1964) NSF 63-53.
No. 26, Library of Congress (April 1964), NSF 64-3.
No. 27, Department of Defense, pt. 2, Defense Documentation Center (October 1964) NSF 64-13.
No. 28, Department of Health, Education, and Welfare, pt. 4, National Library of Medicine (November 1964) NSF 64-23.
No. 29, National Aeronautics and Space Administration (March 1965) NSF 64-29.
No. 30, Department of Agriculture, pt. 2, National Agricultural Library (March 1965) NSF 65-5.
No. 31, U. S. Army, pt. 2, Army Material Command (January 1966) NSF 65-22.
No. 32, Department of Defense, pt. 1, Advanced Research Projects Agency, Defense Atomic Support Agency, Weapons System Evaluation Group, Defense Supply Agency, and 21 Information Analysis Centers (March 1966) NSF 66-9.

Weber, G. A. *The Weather Bureau: Its History, Activities and Organization.* Service Monograph No. 9. Institute for Government Research, New York, 1922, 87 pp.

Wiesner, Jerome B. *Where Science and Politics Meet.* New York: McGraw-Hill 1965. Pages 154 to 162 contain a chapter titled, *What to do About Scientific Information.*

Appendix A: Chronology

1790: First patent act was passed establishing a governmental committee to review and grant patents.

1800: Library of Congress was established.

1807: Coast Survey was authorized and funded by Congress; in 1876 this federal program became the U. S. Coast and Geodetic Survey.

1815: Congress purchased the Thomas Jefferson library to replace the Library of Congress (destroyed when the Capitol Building was burned during the War of 1812); Jefferson's collection was strong in science.

1818: U. S. Army established a medical department that began gathering weather records.

1836: Surgeon General of the Army Joseph Lovell, established a collection of medical books in his office (the forerunner of the National Library of Medicine). Revision of the patent law established the Office of the Commissioner of Patents.

1837: First reported use of engraved printing to produce sailing charts for the U. S. Navy.

1839: Patent Office established an Agriculture Section with a collection of books that became the nucleus for Department of Agriculture Library established in 1863.

1840: Library of Congress was authorized to initiate exchanges with institutions in other countries.

1846: Smithsonian Institution was established with major purpose to "increase and diffuse knowledge among men."

1848: Smithsonian Institution started issuing daily weather maps.

1852: International Exchange Service was organized by the Smithsonian Institution; in 1881 Congress first directly appropriated funds for this service.

1857: Smithsonian Institution weather reports were first printed in the *Washington Evening Star;* National Museum was established in the Smithsonian Institution.

1861: Government Printing Office was established by Congress.

1862: Department of Agriculture was established; enabling legislation included authorization " . . . to acquire and diffuse . . . useful information on subjects connected with agriculture."

1863: Department of Agriculture Library was organized with 1,000 volumes on agricultural subjects transferred from the Patent Office.

172   Appendix A

1865: John Shaw Billings assumed charge of a collection of books (1,800 volumes) known as the Library of the Surgeon General of the Army; in 1871 the collection had increased to 25,000 books and 15,000 pamphlets. Smithsonian Institution began publication of a general catalog of scientific papers; this program was later incorporated into the *International Catalogue of Scientific Papers* that was issued in 250 volumes between 1901 and 1915.

1866: Congress authorized the transfer of the Smithsonian Institution's scientific and technical collection of 40,000 volumes to the Library of Congress.

1868: Department of Agriculture entered into a system of publication exchanges with foreign governments, societies, and individuals.

1869: Brussells Treaty was approved by Congress (the first of a number of international agreements that made the Library of Congress the recipient of official publications of many foreign governments).

1870: Army Signal Corps arranged for simultaneous weather observations at twenty-four dispersed locations.

1873: Smithsonian Institution issued the *Monthly Weather Review* with Cleveland Abbe as Editor; Smithsonian Institution National Weather Service was transferred to Army Signal Corps.

1878: Geological Survey was established in Department of Interior. Coast and Geodetic Survey was established in Treasury Department (an expansion of an earlier federal agency known as the Coast Survey).

1879: First issue of *Index Medicus: A Monthly Classified Record of the Current Medical Literature of the World* was compiled by John Shaw Billings and Robert Flecher. U. S. weather map first published in a newspaper, *New York Graphic.*

1883: Geological Survey began compiling a bibliography of American and foreign geology (the forerunner of the *Bibliography of the Geology of North America*).

1889: Department of Agriculture Library issued the first printed catalog of books in its collections.

1890: National Weather Service was transferred from Army Signal Corps to Department of Agriculture where it became the Weather Bureau.

1898: Department of Agriculture was denied authority to purchase books and periodicals unless appropriations language included authorization for such purposes; the act (30 statutes, 316, March 15, 1898) enabled the departmental library to maintain a record of all departmental library resources.

1899: Department of Agriculture Library initiated printing and distribution of catalog cards of departmental publications.

1900: Department of Agriculture Library began practice of loans of materials to other libraries.

1902: Library of Congress initiated distribution of printed catalog cards.

1903: *Index Medicus* was first published under the patronage of the Carnegie

Institution (one of the earliest cooperative publishing projects in which a government library and a private organization participated).
1911: Department of Agriculture Library began use of photographic copies for interlibrary loan (one of first uses of photographic copies for loan in lieu of originals).
1922: Library of the Surgeon General became the Army Medical Library.
1926: Technical News Service was started (the forerunner of Defense Department sci-tech information activities that later became the Defense Documentation Center).
1934: Department of Agriculture Library cooperated with the American Documentation Institute and Science Service in establishing Bibliofilm; the service provided microfilm copies of publications and manuscripts.
1938: Library of Congress began operation of its photoduplication service; equipment and service was funded with a 1937 grant from the Rockefeller Foundation.
1940: Department of Agriculture initiated program to centralize policy, funding, and operations under the director of libraries in the secretary's office.
1941: Office of Scientific Research and Development was established and fostered the use of sci-tech reports in order to speed dissemination of R&D findings; when it was abolished in 1946, it transferred about 33,000 sci-tech reports to the Office of Naval Research.
1942: *Bibliography of Agriculture* was compiled and published by Department of Agriculture Library.
1943: Army Medical Library supplies microfilm copies of documents without cost to all government agencies and to all individuals connected with accredited institutions in lieu of loan of material.
1945: Air Documents Research Center was established in London to screen "captured" German sci-tech documents. Air Document Division (ADD), Intelligence Department of Headquarters of Air Technical Service, U. S. Air Force was established to catalog, index, reproduce, and distribute copies of captured enemy documents; Technical New Service became a part of this operation. National Advisory Committee for Aeronautics established a sci-tech information program. Office of Declassification and Technical Information was established by Department of Commerce Order No. 5 to be the operating arm of the interdepartmental Publications Board established to prepare plans for declassification of U. S. sci-tech reports and documents according to Executive Order 9569; authority to cover foreign material was included in Executive Order 9604 issued that same year. Manhattan District Editorial Advisory Board was established. Technical Information Section in the Research and Development Division of Oak Ridge National Laboratory was established with a Library Unit that started the formal program for declassification of nuclear science reports and other materials.
1946: Department of Agriculture and American Chemical Society cooperated in

furnishing copies of articles listed in *Chemical Abstracts* to members of the American Chemical Society and subscribers to *Chemical Abstracts;* this program was terminated in 1956 because of possible violation of copyright law. *Bibliography of Scientific and Technical Reports* was first published by Office of Technical Services in the name of the Publications Board. Office of Declassification and Technical Services was renamed Office of Technical Services (OTS) by amendment to Department of Commerce Order No. 5. Office of Naval Research was established and given custody of 33,000 Office of Scientific Research and Development reports. *AEC's Weekly Title List* was started (later became the *Nuclear Science Abstracts*).

1947: Science and Technology Project was started in the Library of Congress to fulfill an Office of Naval Research contract to catalog, index, abstract, and give service on OSRD and other sci-tech reports of interest to the Navy's R&D program. Technical Information Section was established in the Public and Technical Information Division of AEC.

1948: First volume of AEC's *Nuclear Energy Series* was published by McGraw-Hill Book Company. Central Air Documents Office (CADO) replaced Air Documents Division of the U. S. Air Force and was also supported by the Navy's Bureau of Aeronautics; in 1949 the Army joined in its support. Technical Information Section in the Oak Ridge National Laboratory was transferred to the administrative control of Division of Public and Technical Information of AEC. Patent Office initiated a project to develop a mechanized coordinate indexing system to search patents related to the composition of matter. Manhattan District Editorial Advisory Board was replaced by Technical Information Panel to advise AEC on its sci-tech information program; this panel has continued to perform to the present time.

1949: Science Division was established in the Library of Congress; the Library also began to accept contracts from other federal agencies to prepare bibliographies and other information tools on sci-tech subjects and reorganized its Science and Technology Project (renamed the Navy Research Section).

1950: Library of Congress established the Snow, Ice, and Permafrost Establishment project to perform bibliographic services for the Army. American Meteorological Society first published *Meteorological Abstracts* with financial support from the Air Force and with cooperation of Weather Bureau. AEC's Technical Information Section was renamed Technical Information Service (TIS) and the Oak Ridge Technical Information Branch placed under its administration; AEC/TIS established thirty-one regional libraries for its reports, bibliographies, and other publications (selection of the libraries was based upon the recommendations of the American Library Association). Medical Sciences Information Exchange (MSIE) was established in National Research Council to maintain a current inventory of medical sciences research projects supported by the federal government. PL-776 (81st Congress, 2nd sess.) gave Office of Technical Services in Department of Commerce authority to collect and disseminate both foreign and domestic sci-tech information to industry, the government agencies, and the public. John Hopkins Research Project under the direction of Sanford V. Larkey studied problems of medical bibliography with emphasis on the possible application of machine methods for the Army Medical Library. Department of Agriculture contracted with the University of Nebraska to furnish library service to Department employees in the area (the first of several such arrangements).

1951: By executive order, Office of Technical Services became the federal government's sales outlet for federal sci-tech reports. Economic Cooperation Agency, predecessor to the Agency for International Development (AID), signed a contract with Department of Commerce enabling OTS to furnish technical information to Marshall Plan countries in Europe (expanded later to include developing countries). Armed Service Technical Information Agency (ASTIA) was established by Department of Defense; the Joint Defense Research and Development Board was given responsibility for policy guidance of ASTIA, while management control rested with the Research and Development Command of the Air Force.

1952: Office of Scientific Information (OSI) was established in the National Science Foundation. ASTIA assumed control of Central Air Documents Office and renamed it ASTIA Documents Distribution Center. Secretary of Defense Department signed the directive changing the Army Medical Library into the Armed Forces Medical Library, a joint agency of the three military services. Technical Information Division was organized by Library of Congress.

1953: BioSciences Information Exchange was established in the Smithsonian Institution; Medical Sciences Information Exchange was absorbed by the new organization.

1954: Industrial Information Branch was organized in Technical Information Service of AEC to facilitate use by U. S. industry of new technical and scientific findings in nuclear energy.

1955: Patent Office and National Bureau of Standards initiated a cooperative systems development program to design and develop a computer-based information system for Patent Office (called HAYSTAQ).

1956: National Library of Medicine Act was signed into law (PL-941, 84th Congress, 70 Stat. 90); Armed Services Medical Library was transferred to the new organization.

1958: National Aeronautics and Space Agency was established. Baker Panel report to the President's Science Advisory Committee, which was approved by President Eisenhower, urged decentralized approach to sci-tech activities in U. S.; recommended a Science Information Service be established in NSF; and urged a large R&D program in information sciences and technology be mounted by the federal government. Science and Technology Division was formed in the Library of Congress by uniting former Science Division and

Technical Information Division. National Defense Education Act of 1958 (Title IX) authorized a Science Information Service in the National Science Foundation and a Science Information Council to advise the head of this service. National Science Foundation established Office of Science Information Service (OSIS). ASTIA consolidated its Documents Distribution Center and Reference Center in Arlington Hall Station, Va. PL 83-480 (Agriculture Trade Development and Assistance Act) was amended by PL 85-477 that authorized use of surplus foreign credits for translation and acquisition of documents; Executive Order 10900 gave NSF responsibility for coordinating the program. Office of Technical Services and Special Libraries Association agreed to cooperate in preparation of a consolidated list of translations, which was called *Monthly List of Translations*, published by OTS.

1959: Executive Order 10521 was amended by adding Section 10, which said in part that ". . . the National Science Foundation shall provide leadership in effective coordination of scientific information activities of the federal government." Secretary of Agriculture Department placed all departmental libraries outside metropolitan Washington, D. C., under the administrative and budgetary support of the departmental unit they served and thus abolished the centralization program initiated in 1940. National Academy of Sciences/National Research Council established the Office of Documentation. Office of Science and Technology was established in the Executive Office of the President. Federal Council for Science and Technology replaced the former Interdepartmental Committee for Research and Development. National Bureau of Standards established the Research Information Center and Advisory Service on Information Processing (RICASIP). National Science Foundation appointed a Federal Advisory Committee on Scientific Information (abolished in 1961).

1960: Science Information Exchange was established in the Smithsonian Institution through interagency agreement and approval by Office of Science and Technology; BSIE was incorporated in to new organization. NASA established Office of Scientific and Technical Information.

1961: Committee on Scientific Information (COSI) of Federal Council for Science and Technology was established. Federal Council for Science and Technology approved policy of federal agencies honoring page charges by not-for-profit sci-tech journals for articles resulting from federally supported R&D. Subcommittee on Reorganization and Internal Organization of the Senate Committee on Government Operations, chaired by Hubert H. Humphrey, issued report, *Documentation, Indexing and Retrieval of Scientific Information*. Librarian of Congress appointed a committee to study automation for the Library of Congress. National Oceanographic Data Center was organized as part of Navy's Oceanographic Research Center with multiple agency support; this center was transferred to NOAA when it was established.

1962: NASA contracted with American Institute of Aeronautics and

Astronautics to acquire all nonfederal published literature and to catalog index, abstract, and give service on this material, according to NASA specifications. Library of Congress established the National Referral Center for Science and Technology. Department of Agriculture Library was designated by departmental order as the National Agricultural Library. Secretary of agriculture established a committee to plan an information network. President's Special Assistant for Science and Technology, Special Task Force issued report, *Scientific and Technical Communications in the Government* (known as the Crawford Report), which urged R&D agencies to appoint a person to monitor and coordinate sci-tech information activities. ASTIA began supplying twelve OTS Regional Depositories with microfilm copies of unclassified, unrestricted reports. Documentation Incorporated began operation of NASA's computer-based Information Processing Center. Department of Defense established a sci-tech information program through Directive 5100.36.

1963: Director of the Office of Science and Technology appointed a technical assistant for information activities. President's Science Advisory Committee issued Weinberg Report, *Information, Science and Government*, which urged: more active participation of scientists in information activities and assumption of responsibility by federal R&D agencies for effective information services in fields allied to their missions; this latter recommendation became known as the "Delegated Agency" concept. National Bureau of Standards established the National Standard Data Reference System. Ad Hoc Subcommittee (chaired by Roman Pucinski) on National Research Data Processing and Information Retrieval Center of the House Committee on Education and Labor held hearings on HR 1946 and issued the first three parts of report; in 1964 same committee held hearings and issued part 4 of report.

1964: House Select Committee on Government Research (Elliot Committee) issued Study No. IV, *Documentation Dissemination of Research and Development Results* (House Report No. 1932). Stafford Warren, presidential advisor for mentally handicapped, distributed a proposed plan, *National Library of Science System and Network for Published Scientific Literature.* An agreement between Donald Horning, director of the Office of Science and Technology, and Leland Hayworth, director of the National Science Foundation, transferred responsibility for coordination of federal sci-tech information activities from the National Science Foundation to the Office for Science and Technology. COSATI issued its subject category list.

1965: Federal Library Committee was organized with the librarian of Congress as chairman. Office of Science and Technology panel issued report, *Scientific and Technological Communications* (known as the Licklider Report). Medical Library Assistance Act (PL 89-291) was signed into law and became the legal base of NLM's regional medical library network. The National Library of Medicine issued its first "recurring bibliography,"

Index of Rheumatology, and inaugurated its Recurring Demand Searches to provide current awareness service on specific subjects. Office of Technical Service was replaced by Clearinghouse for Federal Scientific and Technical Information (FCSTI). State Technical Services Act (PL 89-182, 79 Stat. 679) was passed; the Department of Commerce organized the Office of State Technical Services to implement the act. ASTIA was expanded to become Defense Documentation Center and was transferred from Defense Research and Engineering to Defense Supply Agency. National Science Foundation, with cooperation from Defense Department and NIH, initiated systems development program to improve discipline-oriented information services; a multi-million dollar, multi-year contract was let to American Chemical Society.

1966: NASA and Army Electronics Research Center began to use DDC's magnetic tapes containing bibliographic data on Defense Department reports. COSATI's National Systems Task Force recommendations were submitted to Federal Council for Science and Technology. NLM established the Toxicological Information Center. Office of Education established the first of its Educational Resource Information Centers (ERIC) as components of its system.

1967: Unesco and ICSU initiated a cooperative project for the study of the feasibility of a flexible network of information systems, known as UNISIST, which resulted in 1971 in Unesco's establishing a program to foster the development of such a world system. Lockheed Corporation installed the Dialog System for experimental use at NASA's Ames Laboratory this on-line interactive system was used by Ames Laboratory scientists and engineers to interrogate NASA's bibliographic data base stored in a computer at Lockheed's Marietta, Georgia, installation. National Library of Medicine's first regional medical library, Countway at Harvard University, was established; NLM organized its own program in biomedical communications; NLM was made a bureau of National Institutes of Health and was also assigned responsibility for the National Medical Audiovisual Center in Atlanta, Georgia.

1968: Education Resources Information Center (ERIC) began to print *Current Index to Journals* in education. National Academy of Sciences/National Research Council, with federal funds, established the Highway Research Information Center as part of the Highway Research Board. World Meteorological Organization organized a "World Weather Watch" in which U. S. participation included an international center located in Washington, D. C.; this center was administrated by the Environmental Sciences Service Agency (ESSA). Lister Hill National Center for Biomedical Communications was established by PL 90-456 and placed in the National Library of Medicine.

1969: National Agricultural Library began distribution of CAIN (Cataloging and Indexing System) tapes, which included data from *Bibliography of Agriculture, Pesticides Documentation Bulletin,* and *NAL Monthly Catalog.*

Department of Defense instituted the policy of requiring that its information Analysis Centers charge for products and services.

1970: National Oceanographic and Atmospheric Agency was formed and immediately established the Environmental Data Service consisting of different information services that became part of the agency when it was formed; later, these and other NOAA information and data services were coordinated into an information system known as OASIS (Oceanic and Atmospheric Scientific Information System). NASA cooperated with AEC/TID in adapting RECON to AEC's operations.

1971: National Science Foundation founded Environmental Information Center at AEC's Oak Ridge National Laboratory. National Technical Information Service (NTIS) replaced the Clearinghouse for Federal Scientific and Technical Information. COSATI was transferred from the Office of Science and Technology to the National Science Foundation; the head of Office of Science Information Service became chairman of COSATI.

Appendix B:
Acronyms and Abbreviations

A&I Services	Abstracting and Indexing Services
ACSTI	Advisory Committee for Scientific and Technical Information of the United Kingdom
AEC	Atomic Energy Commission
AGI	American Geological Institute
AGRI-DOC	An international program for exchange of agriculture documentation (sponsored by FAO)
AIAA	American Institute for Aeronautics and Astronautics
AID	Agency for International Development
ARIST	*Annual Review of Information Science and Technology* (*Annual Review* also used)
ARO	Army Research Office
ASM	American Society for Metals
ASTIA	Armed Services Technical Information Agency
BA	*Biological Abstracts*
BSIE	Biological Sciences Information Exchange or Biosciences Information Exchange
CADO	Central Air Documents Office of the U. S. Air Force
CAN/SDI	Canada Selective Dissemination of Information based in the National Science Library
CAS	Chemical Abstracts Service
CMEA	Council of Mutual Economic Assistance
CFSTI	Clearinghouse for Federal Scientific and Technical Information
CLR	Council of Library Resources
CODATA	Committee for Data for Science and Technology of the International Council of Scientific Unions (also used to refer to the program of this committee and to its operating office)
COSATI	Committee on Scientific and Technical Information of the Federal Council for Science and Technology
COSI	Committee on Scientific Information (the predecessor of COSATI)
CRDSD	*Current Research and Development in Scientific Documentation* (publication issued by NSF)
DDS	Defense Documentation Center
DOD	Department of Defense

DOT	Department of Transportation
DSIR	Department of Scientific and Industrial Research of the United Kingdom
ECA	Economic Cooperation Administration of the Department of State
EJC	Engineers Joint Council
ERIC	Educational Research Information Center (later changed to Educational Resources Information Center)
ESRO	European Space Research Organization
ESSA	Environmental Sciences Service Agency
EURATOM	European Atomic Energy Agency
FACSI	Federal Advisory Committee for Scientific Information (reported to the director of NSF)
FCST	Federal Council for Science and Technology
GRACE	Graphic Arts Composing Equipment (developed for NLM)
GPO	Government Printing Office
GSA	Geological Society of America
GSIS	Group for Standardization of Information Services
HAYSTAQ	Meaningless acronym used to identify a cooperative systems development project of National Bureau of Standards and Patent Office
HUD	Department of Housing and Urban Development
IAEA	International Atomic Energy Agency
ICSU	International Council of Scientific Unions
ICSUAB	International Council of Scientific Unions Abstracting Board
ICIREPAT	International Cooperation in Information Retrieval Among Examining Patent Offices
IGO	International governmental organization
IGRIS	Interagency Group for Research on Information Systems
IMC	International Meteorological Committee
IMO	International Meteorological Organization
INIS	International Nuclear Information Service
IPG	Information Policy Group of the OECD
ISB	International Standard Book numbering system
ISI	Institute for Scientific Information
ISSN	International Standard Serials Numbering System
JICST	Japan Information Center for Science and Technology
LC	Library of Congress
MARC	Machine Readable Catalog (system develop by the Library of Congress)
MDEAB	Manhattan District Editorial Advisory Board
MEDLARS	Medical Literature Analysis and Retrieval System of the National Library of Medicine

Acronyms and Abbreviations   183

MESH	Medical Subject Headings of the NLM
MIT	Massachusetts Institute of Technology
MSIE	Medical Sciences Information Exchange
NACA	National Advisory Committee on Aeronautics
NAE	National Academy of Engineering
NASA	National Aeronautics and Space Agency
NAS/NRC	National Academy of Science/National Research Council
NAL	National Agricultural Library
NATO	North Atlantic Treaty Organization
NGO	Nongovernmental international organization
NIH	National Institutes of Health
NLM	National Library of Medicine
NOAA	National Oceanographic and Atmospheric Agency
NRC	National Research Council (used for institutions located in Canada and the United States)
NRS	Navy Research Section of the Library of Congress
NSA	Nuclear Science Abstracts
NSDRS	National Standard Data Reference System
NSL	National Science Library (Canada)
NTIS	National Technical Information Service
OECD	Organization for Economic Cooperation and Development
ONR	Office of Naval Research
OSIS	Office of Science Information Service in the National Science Foundation
OSRD	Office of Scientific Research and Development
OST	Office of Science and Technology in the Executive Office of the President
OSTI	Office of Scientific and Technical Information of the United Kingdom
OSTS	Office of State Technical Services in the Department of Commerce
OTS	Office of Technical Services in the Department of Commerce
PSAC	President's Science Advisory Committee
R&D	research and development
RANN	Research Addressed to National Needs (a NSF directorate)
RECON	REmote CONsole, (NASA's on-line information system)
RESPONSA	computer-based information service of AEC
SATCOM	Committee on Scientific and Technical Communications (National Academy of Science/National Research Council)
SCAN	selective dissemination system (Office of Scientific and Technical Information, NASA)
sci-tech	scientific and technical
SDC	Systems Development Corporation

SEATO	Southeast Asia Treaty Organization
SIPRE	Snow, Ice, and Permafrost Establishment of U. S. Army Corps of Engineers (also used for a project in the Library of Congress)
SLA	Special Libraries Association
SSIE	Smithsonian Science Information Exchange
STAR	*Space Technology and Research* (abstract bulletin of NASA)
STP	Science and Technology Project of the Library of Congress
TAB	*Technical Abstract Bulletin* (Department of Defense)
TID	Technical Information Division of the AEC formerly Technical Information Section (TIS)
TIP	Technical Information Panel (advisory committee on scientific and technical information for the AEC)
TIS	Technical Information Section of the AEC (later TIS referred to the same organization but meant Technical Information Service); also Technical Information Service of Canada
UNIDO	United Nations Industrial Development Organization
USDA	United States Department of Agriculture
USGS	United States Geological Survey
VINITI	All-Union Institute for Scientific and Technical Information of the USSR
UNISIST	Meaningless acronym used to identify a world information network being sponsored by Unesco and ICSU
Unesco	United Nation Educational, Scientific, and Cultural Organization
WHO	World Health Organization of the United Nations
WMO	World Meteorological Organization of the United Nations

Appendix C:
White House Press Release Containing Baker Report

The President today approved a plan designed to help meet the critical needs of the Nation's scientists and engineers for better access to the rapidly mounting volume of scientific publication.

Acting upon the recommendations of his Science Advisory Committee, the President directed that the National Science Foundation take the leadership in bringing about effective coordination of the various scientific information activities within the Federal Government. The President asked that all Federal agencies whose programs involve scientific information cooperate with and assist the National Science Foundation in improving the Government's own efforts in this area.

Today's action by the President strengthens and reinforces the provision of the "National Defense Education Act of 1958" calling for the establishment of a Science Information Service in the National Science Foundation to: "Provide or arrange for the provision of, indexing, abstracting, translation, and other services leading to a more effective dissemination of scientific information, and undertake programs to develop new or improved methods, including mechanized systems for making scientific information available."

The Committee urged that fullest use be made of existing information services, both public and private, and that the Foundation's Science Information Service supplement rather than supplant present efforts.

Dr. James R. Killian, Jr., Special Assistant to the President for Science and Technology and Chairman of the Science Advisory Committee, commented on the growing dimensions of world scientific publication to the extent that it has become a problem requiring action at the national level.

"Science and engineering are largely built on the published record of earlier work done throughout the world," Dr. Killian stated. "There are, for example, 55,000 journals appearing annually, containing about

---

See Anne Wheaton, Associate Press Secretary to the President, Press Release for A.M. Papers, Sunday, December 7, 1958.

1,200,000 articles of significance for some branch of research and
engineering in the physical and life sciences. More than 60,000 different books are published annually in these fields, while approximately 100,000 research reports remain outside the normal channels of
publication and cataloging. Within this vast body of world-wide scientific information, published and unpublished, lie the technical data
that scientists need in order to do their work. The situation is
further complicated by the fact that a large and important proportion
of the world's scientific literature appears in languages unknown to
the majority of American scientists, such as Russian and Japanese."

In its recommendations, the President's Science Advisory Committee
outlined a program calling for the review, coordination, and stimulation, on a nation-wide basis, of activities in the areas of primary
and secondary publications, scientific data centers, unpublished research information, storage and retrieval, and translation by mechanical
means.

No new agency will be required to carry out the recommended program.
Under its enabling act, the National Science Foundation has devoted
special attention to the scientific information needs of scientists
and has developed a series of programs designed to help meet those
needs. At least ten other Federal agencies are engaged in abstracting
and indexing, translating, preparation of technical reports, and research related to information needs. These agencies are asked to
cooperate in providing or arranging for acquisition and reference programs, clearinghouse functions, and evaluation studies of existing
programs. Research on new and improved methods of information handling
will be emphasized and the Department of State will take the lead in
encouraging cooperation among the United States, foreign and international scientific information organizations.

The President's Science Advisory Committee considered the whole problem
of such importance that earlier this year it appointed a special subcommittee to consider the subject at length. Headed by Dr. W. O. Baker,
Vice-President (Research), Bell Telephone Laboratories, the subcommittee
comprises the following members: Mr. Curtis Benjamin, President, McGraw
Hill Book Company; Dr. Caryl P. Haskins, President, Carnegie Institution
of Washington; Dr. Elmer Hutchisson, Director, American Institute of
Physics; Dr. Warren C. Johnson, Dean, Division of Physical Sciences,
U. of Chicago; Mr. Don K. Price, Dean of the School of Public Administration and Littauer Professor, Harvard University; Dr. H. Scoville; Dr.
Alan T. Waterman, Director, National Science Foundation.

In submitting its findings the subcommittee paid special tribute to
the work of individual scientists and engineers in selecting, interpreting, and abstracting scientific and technical information. It
noted the fact that the services rendered by many of the scientific
societies and professional institutions to the scientific community in
the information field are world famous for their quality. It expressed
the hope that such private groups would continue to cooperate with and
assist the Federal Government in the achievement of long-range solutions
to scientific information problems.

The subcommittee's conclusions form the basis for the recommendations
submitted to the President by the Science Advisory Committee.

A REPORT OF
THE PRESIDENT'S SCIENCE ADVISORY COMMITTEE

IMPROVING THE AVAILABILITY
OF SCIENTIFIC AND TECHNICAL
INFORMATION IN THE UNITED STATES

*****

WHAT THE PROBLEM IS AND WHY IT IS SERIOUS

The long, hard look we have recently taken at the state of science and technology in this country has brought to light several areas that need to be strengthened and improved. Some of these, notably in the field of education, have aroused nation-wide concern. But another area--also in great need of attention--has attracted little or no public interest. This is the matter of scientific information--the technical data that a scientist needs in order to do his job. Yet our progress in science may very well depend upon the intelligent solution of problems in that area.

All of us use a wide variety of information every day of our lives. We glean it from newspapers, conversation, radio and television, magazines, clocks, books, meters, mail, maps and so on. The scientist, however, is interested in the specialized information that results from scientific research. The publication of research information that results from scientific research. The publication of research information is absolutely essential to every working scientist for two reasons: (1) It is the means by which he announces significant results in his own work, establishes priority where appropriate and invites the evaluation of other scientists; (2) It is also the means by which he keeps abreast of what others are doing in his field.

The extent to which the working scientist depends upon the work of others has been clearly stated by one of the greatest of all scientists, the atomic physicist, Ernest Rutherford. As quoted by James Newman in a recent issue of The Scientific American, Lord Rutherford said:

> I have also tried to show you that it is not in the nature of things for any one man to make a sudden violent discovery; science goes step by step, and every man depends on the work of his predecessors. When you hear of a sudden unexpected discovery-- a bolt from the blue as it were--you can always be sure that it has grown up by the influence of one man on another, and it is this mutual influence which makes the enormous possibility of scientific advance. Scientists are not dependent on the ideas of a single man, but on the combined wisdom of thousands of men, all thinking the same problem, and each doing his little bit to add to the great structure of knowledge which is gradually being erected.

The reason scientific information has become a major problem, particularly since World War II, is that the rapid rate of scientific progress has multiplied the volume of scientific information to a point where it can no longer be published and handled within the framework of existing methods. When one considers, too, that much of what is significant in science is being published in unfamiliar languages, it is clear that the working scientist is faced with almost insuperable problems in attempting to keep himself informed on what he needs to know.

Some idea of the volume of increase may be had from the fact that the science and technology periodical collections of the Library of Congress have doubled approximately every 20 years for the past century and now contain approximately a million and a half volumes, a significant fraction of the Library's total bound collections. The Library is receiving journals in science and technology at the rate of about 15,000 annually, and 1,200 to 1,500 new periodicals are appearing each year. Yet the Library receives less than a third of the 50,000 scientific periodicals that appear in the world list of 1952 and it is expected that by 1979 the total world output will reach 100,000 journals.

The language difficulty is reflected in the fact that Russian-language publications are estimated to account for a tenth or more of all the scientific literature being published in the world today. This Russian total is second only to English.

Reduced to simple terms, the scientist's problem with respect to information is: How can the present volume of research results be published promptly? What is being published now? Where is it? and How can I get at it? The purpose of this paper is to examine these problems and to suggest possible ways in which they can be solved. In particular, it will consider the question of what should be the responsibility of the Federal Government in meeting this crisis.

THE PRESENT SYSTEM

The system by which scientific information is disseminated is the result of evolution rather than any preconceived system or plan. Its defects stem largely from its inability to keep pace with the increasing volume of scientific results and literature and the absence of techniques geared to the newer forms of scientific information, such as Government reports. The situation is further complicated by the fact that a large and important proportion of the world's scientific literature appears in languages unknown to the majority of American scientists, such as Russian and Japanese.

Scientific information appears in several forms. Most significant are the highly specialized technical periodicals, called primary journals, because it is in these that new scientific results are first published. The Physical Review, Journal of the American Chemical Society, and the Aeronautical Engineering Review are examples.

Another important primary source is the monograph, an exhaustive study of some highly specialized phase of science. Because it is of interest to only a limited number of scientists, and because it often includes elaborate charts and plates, the monograph is almost prohibitively expensive to publish. The result is a lack in this country of monographs on many exceptionally important scientific subjects that should be so covered.

A second important category is the abstracting journals, such as Biological Abstracts and Chemical Abstracts. These contain summaries of synopses of papers that originally appeared in primary journals. When adequately indexed, they permit a searcher to locate previously published papers on any given subject. If an abstract is sufficiently informative, it may serve the scientist in lieu of the complete paper. It should be noted parenthetically, however, that the 14 major scientific abstracting services in the United States recently indicated that the almost half a million abstracts that they issue annually constitute only about 55 per cent of what they should be publishing in order to cover the literature in their combined fields reasonably well. Other important secondary sources include critical reviews, special indexes and indexing services, bibliographies, title lists, collected tables of contents, handbooks of data, and compendia of various kinds.

A recent trend of special interest is the establishment of Data Centers. When the quantity of research data in a given field becomes too great for book publication to be practical, the Data Center offers a solution. Such centers compile, correlate, standardize, and organize numerically, data representing the properties of materials or the characteristics of phenomena. Examples of such centers include the Thermophysical Properties Research Center at Purdue University; American Petroleum Institute Research Project 44 at the Carnegie Institute of Technology, which is concerned with the physical properties of hydrocarbons; the Nuclear Data Project of the National Research Council; and the National Bureau of Standards center on Selected Values of Chemical Thermodynamic Properties.

Falling outside scientific information that is published, cataloged, and indexed in the normal way, is a steadily mounting volume of Government research reports. It is conservatively estimated that upwards of 50,000 scientific reports (at least half of which bear no security classification) are issued annually by the private and Government laboratories that conduct Federally-sponsored research. Many of the newest and most significant scientific data are to be found in these reports.

A smaller body of scientific information not covered by the normal processes is to be found in such material as research findings submitted in satisfaction of Ph.D thesis requirements, industrial reports and papers presented at scientific meetings and symposia.

At the present time it is not even possible to answer the question with any degree of completeness, "What is being published now?" One would assume that, somewhere in the world, there must be a composite listing of the world's scientific publications--perhaps even arranged by subject fields--but no such compilation exists. The establishment of such a list and its maintenance on a current basis obviously would be a very expensive undertaking, and this is one reason why it has never been done.

The basic answer to "Where can I find it?"--as far as journals are concerned--is the "Union List of Serials," in the libraries of the United States and Canada. Such a compilation lists periodicals alphabetically and names the libraries where each can be found. But no such union list of scientific journals now exists. A Joint Committee on a Union List of Serials covering all fields has estimated that the science and technology portion of a new union list would cost approximately three-quarters of a million dollars. It could be kept up to date only in a relative sense, since such a list is constantly changing. It follows, of course, that no comprehensive listing of the principal secondary publications is in existence either.

Then there is the problem of "How can I get it?" The scientist who needs a particular journal may find himself (if the journal is rare) far distant from the location of the nearest copy as indicated by the union list; or he may find that the article he is seeking is in a language he does not read.

In summation, then, it may be said that both inside and outside the normal channels of scientific communication a mounting flood of scientific data threatens to swamp even the most zealous research investigator. The implications go far beyond the inability of one man, or even a group of men, to keep abreast of developments in their field. Our very progress in science is dependent upon the free flow of scientific information, for the rate of scientific advance is determined in large measure by the speed with which research findings are disseminated among scientists who can use them in further research.

HOW ARE WE GOING TO MEET THIS PROBLEM?

The situation has evolved over a lengthy period of time, during which the developing problems not only have been recognized, but have been the subject of attack on a number of separate fronts. These efforts have been handicapped, however, by the lack of overall coordination and sufficient funds with which to support really effective remedies.

What is Already Being Done?

All along the line there have been sincere efforts to cope with the problems. Primary journals have expanded substantially in recent years and the scientific societies have helped to cover the increased costs by raising dues and subscription prices. In an effort to conserve space, greater and greater condensation of papers is being required, with the result that there is danger of few people besides the author and his immediate colleagues being able to understand a paper. There is a constant search for cheaper production methods and many journals levy page costs upon the authors, so that scientists must pay for the privilege of having their research findings published. Such financial help as the Government has given has been limited, consisting largely of short-term emergency grants made to tide a particular journal over a rough spot or to launch a new journal that is badly needed in order to fill a gap. Some agencies pay page costs for their employees and their contractors' employees when they publish.

Federal aid has also been provided in the form of temporary assistance to commercial abstracting and indexing services, including funds to support the establishment of a National Federation of Science Abstracting and Indexing Services, designed to bring cooperative efforts to bear upon mutual problems. A few Government agencies publish or

partially support certain secondary publications in subject fields of
particular interest to them.

It is generally agreed, however, that the magnitude and seriousness of the problem are such that a long-term solution requires fundamental research into the problem and widespread application of machine methods and techniques. In other words, science must look within itself for a new system that will meet present-day requirements for the location, storage, and retrieval of scientific information.

A number of industrial firms have developed, and are using successfully, mechanized storage and retrieval systems tailored to their own needs. Large manufacturers of business machines and computers are becoming increasingly interested in the application of their equipment to information-processing problems. A dozen or more universities are carrying on research in the information-handling field, including studies of existing patterns of scientific communication in various subject fields, research in mechanical translation, development of procedures for determining how scientists use technical information, and research on actual mechanical systems for information storage and retrieval. Within the Government, the National Science Foundation has supported research on scientific information problems to the extent that available funds have permitted.

Efforts are also being made to improve the availability of foreign scientific information. The emphasis is on Russian research results because Soviet scientific publications are second only to our own in number, and because so few scientists in this country read Russian. Of the 61 Soviet journals available here on subscription in cover-to-cover translation, about 34 are being supported principally by the National Science Foundation, with assistance from the Atomic Energy Commission and the Office of Naval Research. Nine are supported by the National Institutes of Health; the rest are issued commercially.

In the field of unpublished documents the Office of Technical Services, Department of Commerce, lists some 7,500 such documents each year in its abstracting journal, U. S. Government Research Reports. Copies of all items so announced can be obtained in original form or in photoreproduction. The Library of Congress is building in its Science and Technology Division an open reference collection of unclassified reports. The National Science Foundation maintains a clearinghouse for Government research information to provide scientists information on Government-supported research in their fields and the reports that are available.

Thus a considerable amount of work is being done on serious scientific information problems. From the standpoint of national welfare, however, these efforts are on far too small a scale to deal with the overall problem. The question then remains as to how it can be met.

What Should be done for the Future?

Two alternative possibilities have been advanced. One would be the establishment of a large and highly centralized scientific information agency, financed by the Federal Government or by government and

private industry. A second would be the establishment of a science
information service of the coordinating type, which would strengthen
and improve the present system by taking full advantage of existing
organizations and the specialized skills of persons with long experience in the field. Let us examine the respective merits of these
alternatives.

<u>A Single Large Operating Center</u>? The proposal to solve existing
problems in the field of scientific information by the establishment
of a single large operating center, financed wholly or in part by the
Federal Government, may have been suggested by the experience of the
Soviet Union with its All-Union Institute of Scientific Information.
The organization and operation of the Institute implies that the
Russians recognize the magnitude and importance of the problem by their
decisive and aggressive attempts to meet it. Available evidence indicates that the Institute operates effectively in meeting the needs of
Russian science. But, it must not be overlooked that in planning the
establishment and operations of the Institute, the Russians could not
call upon the services of scientific information organizations such as
we find already in existence in the private enterprise structure of
our country, and which have been in operation many years.

The solution the Russians have developed for meeting their own
problems in our judgment <u>would</u> <u>not</u> be equally effective in meeting
ours. The Russian Institute is organized along the lines that are
basically compatible with the organization and administration of research in the Soviet Union, which, of course, is controlled by the
Central Government. Our own research efforts are organized and administered very differently, and it is illogical to suppose that a
highly centralized organization for the dissemination of research information would serve our purposes equally well. Whatever its faults
may be, our present system has developed along the lines of individual
initiative and private enterprise that are very basic to our institutions.

The primary journals, as well as the abstracting services, are
published under the benign auspices of the scientific societies who
are in a better position than anyone else to appreciate the information problems of scientists. Existing services, moreover, represent
a considerable investment of private capital. <u>Chemical Abstracts</u>,
for example, which has operated without Government subsidy, had a 1957
budget of approximately $1.5 million. Although most of the journals
and services have smaller budgets and many do receive some Government
support, the total private investment in the publication and dissemination of results of scientific research runs into many millions of
dollars. The mere mechanics of transforming the existing decentralized
system of private enterprise into a strong central agency are enough
to stagger the imagination.

From a purely practical point of view, it must be remembered that
much of the day-to-day work involved in the dissemination of scienti-
fice information--that is, the writing, editing, abstracting, translating, and so on--is done either by scientists or people with technical skills of a very high order. Many of these people perform such
chores in addition to their regular scientific work and it is quite
inconceivable that they could be induced to affiliate themselves on a

full-time basis with a centralized agency. Put the matter another way: The case for a Government-operated, highly centralized type of center can be no better defended for scientific information services than it could be for automobile agencies, delicatessens, or barber shops.

A <u>Science</u> <u>Information</u> <u>Service</u>? The second alternative, however, could lead to an integrated, efficient and comprehensive scientific information service that would take advantage of privately supported programs as well as the very extensive work being done by the Federal Agencies--that is, it would strengthen rather than supplant them. Specifically, this solution calls for the establishment within the Government of an organization that might be called a Science Information Service. Such a Service would assist, cooperate with, and supplement the many existing scientific information programs but would "take over" none of them. It would retain the benefits of the existing complex of scientific information services while working at the same time toward remedying its defects. Such a program would be in the best American tradition of private enterprise and Government working together voluntarily for the national good.

The Service would have two important functions: (1) through effective coordination and cooperative effort of public agencies and private organizations to capitalize upon and improve existing facilities and techniques in such a way as to afford immediate relief to short-term problems of a pressing nature; and (2) to encourage and support a fundamental, long-term program of research and development, looking to the application of modern scientific knowledge to the overall problem through the application of machine techniques and through yet-undiscovered methods.

Under the first category the Service would help to answer the scientist's fundamental questions: How can the present volume of research results be published promptly? What is being published now? Where is it? and How can I get it?

In the area of primary publication, the Service would provide financial assistance where needed for the publication of journals and monographs. It would encourage publishers and scientific societies to experiment with new and streamlined methods of publication designed to increase efficiency, improve services, and decrease costs. Similar cooperation would be encouraged among the producers of secondary publications, and financial assistance provided when necessary.

The Service would provide the answer to "What is being published now?" by sponsoring, and if necessary supporting, the immediate preparation of world-wide lists of both primary and secondary scientific research publications, subject-classified and indexed. It would perform a similar task with reference to a union list of scientific and technical periodicals and provide a clearinghouse of information on abstracting and indexing services throughout the world. It would review the newly developing field of Data Centers, compiling information on those that now exist, analyzing overlaps and duplications, and defining areas where new centers are needed.

The whole area of foreign scientific information would be scrutinized and the translation of Russian science expanded to the extent needed to provide full coverage. Additional translation programs in Japanese and other languages would be initiated as needed.

The Service would give special attention to the area of Government scientific reports by expanding the existing announcement system to include every significant unclassified report. It would also expand and improve facilities for making copies of these reports available upon request. It would foster cooperative projects among the agencies to promote greater efficiency in the preparation, processing, and dissemination of Government reports.

It would seek to expand and improve inter-library exchange agreements throughout the world, photocopying processes, and other ways and means of bringing to the scientist copies of items unattainable through normal channels.

All of these things, the Service, with sufficient funds and backing, could proceed to do at once. For the longer term, the Service should support a continuing program of research and development through grants and contracts, looking to the widespread application of machine techniques to such problems as storage, retrieval, indexing, and on a higher plane, to such problems as translation and abstracting.

CONCLUSION

It is clear that in the realm of scientific information, the scientist has neglected his own needs. As a nation we have readily applied modern scientific knowledge to the solution of much more difficult problems. If the Federal Government will establish a national coordinating service of the type that has been described, we can move toward solution of a problem that is vital to our progress in science.

Fortunately a new agency will not be required to meet this need. The National Science Foundation, whose enabling Act charges it with specific responsibilities for scientific information, already has a pilot program in this field and hence useful experience and special competence. The Foundation plays a coordinating role with respect to basic research and policy matters with the Federal Government. The establishment of the Science Information Service within the Foundation could be easily achieved by the extension of the Foundation's present program.

The Committee therefore recommends that the National Science Foundation expand its scientific information program to consititute a Science Information Service that would serve to aid and coordinate existing governmental and private efforts.

Dr. James R. Killian, Jr., Chairman	Dr. George B. Kistiakowsky
Dr. Robert F. Bacher	Dr. Edwin H. Land
Dr. William O. Baker	Dr. Edward M. Purcell
Dr. Lloyd V. Berkner	Dr. Isidor I. Rabi
Dr. Hans A. Bethe	Dr. H. P. Robertson
Dr. Detlev W. Bronk	Dr. Paul A. Weiss
Dr. James H. Doolittle	Dr. Jerome B. Wiesner
Dr. James B. Fisk	Dr. Herbert York
Dr. Caryl P. Haskins	Dr. Jerrold R. Zacharias

MEMBERSHIP
OF
THE PRESIDENT'S
SCIENCE ADVISORY COMMITTEE

DR. JAMES R. KILLIAN, JR., Chairman, Special Assistant to the President for Science and Technology, The White House

DR. ROBERT F. BACHER, Professor of Physics, California Institute of Technology

DR. WILLIAM O. BAKER, Vice President (Research), Bell Telephone Laboratories

DR. LLOYD V. BERKNER, President, Associated Universities, Inc.

DR. HANS A. BETHE, Professor of Physics, Cornell University

DR. DETLEV W. BRONK, President, Rockefeller Institute for Medical Sciences and President, National Academy of Sciences

DR. JAMES H. DOOLITTLE, Vice President, Shell Oil Company

DR. JAMES B. FISK, Executive Vice President, Bell Telephone Laboratories

DR. CARYL P. HASKINS, President, Carnegie Institution of Washington

DR. GEORGE B. KISTIAKOWSKY, Professor of Chemistry, Harvard University

DR. EDWIN H. LAND, President, Polaroid Corporation

DR. EDWARD M. PURCELL, Professor of Physics and Nobel Laureate, Harvard University

DR. ISIDOR I. RABI, Professor of Physics and Nobel Laureate, Columbia University

DR. H. P. ROBERTSON, Professor of Physics, California Institute of Technology

DR. JEROME B. WIESNER, Director, Research Laboratory of Electronics, Massachusetts Institute of Technology

DR. HERBERT YORK, Chief Scientist, Advanced Research Projects Agency, Department of Defense

DR. JERROLD R. ZACHARIAS, Professor of Physics, Massachusetts Institute of Technology

DR. PAUL A. WEISS, Rockefeller Institute for Medical Science

Appendix D: Excerpts from the Crawford Report

The following are the ten major recommendations of the Baker task force report to the President's Special Assistant for Science and Technology. The statement preceding the recommendations stimulated the appointment of a special assistant in the Office of Science and Technology who primarily was responsible for coordination and planning of federal sci-tech information activities. Between 1962 and 1970 all the recommendations included in this report were partially or fully implemented by the executive branch departments and agencies.

GOVERNMENT-WIDE POLICY DIRECTIONS AND REVIEW

We believe that Government-wide policy direction and review of Federal STINFO activities should not be assigned to any one of the existing Departments or Independent Agencies. There are natural difficulties involved whenever any one agency attempts to direct and review activities of another. These difficulties would be compounded if the function were assigned to any agency which participates in any way in the communication of STINFO, since it would be placed in the awkward position of having to judge objectively its own performance. The dictates of good management alone would discourage any such dual responsibility and involvement.

In our opinion this broad functional responsibility is best located in the Executive Office of the President on a permanent basis. The President's recent Reorganization Plan No. 2, establishing an Office of Science and Technology in the Executive Office of the President, provides a logical placement of the Government-wide policy direction and review function.

See James H. Crawford et al., Scientific and Technical Communications in the Government: Task Force Report to the President's Special Assistant for Science and Technology (Springfield, Va.,: Clearinghouse for Scientific and Technical Information, 1962), pp. 81.

Appendix D

RECOMMENDATIONS

1. There should be promulgated to the Departments and Independent Agencies of the Federal Government an official statement of the President which gives formal, high-level recognition to the urgent and important character of STINFO communication in relation to Federally-supported R&D programs and which announces the official common purpose of such Federal STINFO activities.

2. There should be established within the structure of the Executive Branch of the Government an organization structure focal point of responsibility for the Government-wide direction and review of Federal Government programs and activities for the communication of scientific and technological information.

3. Each R&D agency of the Federal Government should be directed to establish internally an appropriate, formal organizational focus of responsibility and authority for agency-wide direction and control of STINFO activities.

4. In collaboration with its agency counterparts, the recommended focal point for Government-wide direction and review should develop and announce a logical structure of defined functions which will guide all activities involved in the management and operation of the Federal Government's STINFO system.

5. Government- and agency-wide points of responsibility, in close cooperation, should also develop and guide the application of realistic criteria to govern the functional operations of the communication process:...

6. There should be established within the structure of the Executive Branch a Government-wide clearinghouse capability for information regarding currently planned and active research and development efforts.

7. There should be established within the structure of the Executive Branch a Government-wide clearinghouse capability for documents reporting the results of research and development work supported by the Federal Government.

8. There should be established in the structure of the Executive Branch a Government-wide clearinghouse capability for retrospective search and retrieval services of Federally-supported, organized collections of scientific and technological information.

9. There should be established within the structure of the Executive Branch a Government-wide clearinghouse capability for coordinated access to Federally-supported specialized information centers and services.

10. There should be established within the structure of the Executive Branch a Government-wide clearinghouse capability of information related to formal scientific and technical meetings supported by the Federal Government.

Appendix E: Excerpts from the Weinberg Report

See President's Science Advisory Committee, *Science, Government and Information: The Responsibilities of the Technical Community and the Government in the Transfer of Information* (Washington, D. C.: Government Printing Office, 1963), pp. 52.

SUMMARY AND MAJOR RECOMMENDATIONS

Transfer of information is an inseparable part of research and development. All those concerned with research and development—individual scientists and engineers, industrial and academic research establishments, technical societies, Government agencies—must accept responsibility for the transfer of information in the same degree and spirit that they accept responsibility for research and development itself.

The later steps in the information transfer process, such as retrieval, are strongly affected by the attitudes and practices of the originators of scientific information. The working scientist must therefore share many of the burdens that have traditionally been carried by the professional documentalist. The technical community generally must devote a larger share than heretofore of its time and resources to the discriminating management of the ever-increasing technical record. Doing less will lead to fragmented and ineffective science and technology.

These are the major findings and recommendations of this Panel. In arriving at these conclusions, the Panel has tried to understand the information transfer process itself, and to identify those problems in information handling that have been magnified by the accelerating growth of science and technology. The first two parts of the following report therefore describe some attributes of the information process and of various information handling systems.

Since strong science and technology is a national necessity, and adequate communication is a prerequisite for strong science and technology, the health of the technical communication system must be a concern of Government. Moreover, since the internal agency information systems overlap with the non-Government systems, the Government must pay attention to the latter as well as to the former.

The Government must be concerned with our non-Government communication systems for another, less obvious reason. The technical literature with its long tradition of self-criticism helps, by its very existence, to maintain the standards, and hence the validity, of science, particularly of basic science. The Government, as the largest supporter of basic science, has a strong interest in keeping viable this mechanism of critical review of the science it supports.

The Government's concern with technical communication is complicated by the impact of modern science and technology on national defense. Criteria for guarding information that should not be divulged in the national interest must be established and must be kept up to date. This Panel has not analyzed in detail these difficult problems of secrecy and classification; they may well bear further thought and analysis by another group.

Since both the Government and the technical community are involved with our technical communication system, the Panel, in making detailed recommendations that elaborate upon our general recommendations, has addressed itself both to the technical community and to the Federal agencies.

A. *Recommendations to the Technical Community*

1. *The technical community must recognize that handling of technical information is a worthy and integral part of science (pp. 14, 27, 29).*

We shall cope with the information explosion, in the long run, only if some scientists and engineers are prepared to commit themselves deeply to the job of sifting, reviewing, and synthesizing information; i.e., to handling information with sophistication and meaning, not merely mechanically. Such scientists must create new science, not just shuffle documents: their activities of reviewing, writing books, criticizing, and synthesizing are as much a part of science as is traditional research. We urge the technical community to accord such individuals the esteem that matches the importance of their jobs and to reward them well for their efforts.

2. *The individual author must accept more responsibility for subsequent retrieval of what is published (pp. 14, 24–26).*

Individual scientists and engineers must participate in the information transfer process, rather than leaving the entire responsibilty to the professional documentalist. We therefore urge authors of technical papers to—

   a. Title papers in a meaty and informative manner (p. 24)

   b. Index their contributions with keywords taken from standard thesauri. Societies and editors are urged to establish such thesauri wherever this is practical (p. 25).

   c. Write informative abstracts (p. 25).

   d. Refrain from unnecessary publication (pp. 25–26).

3. *Techniques of handling information must be widely taught (p. 28).*

Familiarity with modern techniques of information processing is necessary for the modern scientist and engineer. Our colleges and universities must provide instruction in these techniques as part of the regular scientific curriculum. They must also educate in the art of handling information more professionals who can lighten the burden of the technical man and can invent new techniques of information retrieval.

4. *The technical community must explore and exploit new switching methods (p. 30).*

The information transfer network is held together by an array of switching devices that connect the user with the information (as contrasted with the documents) he needs. As the amount of information grows, more ingenuity will be needed to find effective switching mechanisms, if only because the capacity of the human mind places a limit on how much information can be assimilated. The technical community must courageously explore new modes for information processing and retrieval. Among the schemes that ought to be exploited more fully are:

a. Specialized Information Centers (pp. 14, 32–33, 43). The Panel sees the specialized information center as a major key to the rationalization of our information system. Ultimately we believe the specialized center will become the accepted retailer of information, switching, interpreting, and otherwise processing information from the large wholesale depositories and archival journals to the individual user. The Panel therefore urges that more and better specialized centers be established.

We believe the specialized information center should be primarily a technical institute rather than a technical library. It must be led by professional working scientists and engineers who maintain the closest contact with their technical professions and who, by being near the data, can make new syntheses that are denied those who do not have all the data at their fingertips. Information centers ought to be set up where science and technology flourish. We believe that the large, Government-supported laboratories could become congenial homes for groups of related specialized information centers.

b. Central Depositories (pp. 30–32). The central depository to which authors submit manuscripts that are announced and then distributed on request may ease the technical problems of switching documents quickly and discriminatingly between user (particularly the specialized center) and source. Central depositories are now being used by several Government information systems, and there is little question of their practicality. The Panel, though recognizing the difficulties of replacing the traditional techniques of communication via conventional journals, nevertheless urges technical societies to experiment with central depositories, or some variant thereof (as is done by the American Physical Society), for at least some of their literature.

c. Mechanized Information Processing (pp. 20–21, 34–35). The Panel recognizes that mechanical equipment offers hope for easing the information problem. Commercially available equipment is not the remedy in every case; economics, size, frequency of use, growth rate, depth and sophistication of indexing must be examined in detail for each collection before a specific system is to be mechanized. There is a need for equipment specifically designed to retrieve documents from very large collections. The recent study under the auspices of the Council of Library Resources, recommending automation of the Library of Congress, should be evaluated with a view toward its implementation both as a means of improving the services offered by the Library and of advancing the art of automatic retrieval.

d. Development of Software (p. 35). Hardware alone is not a panacea for difficulties of information retrieval. Software, including methods of analyzing, indexing, and programming, is at least as necessary for successful information retrieval. The Panel wishes to call the attention of the technical community to a promising new method of access to the literature called the citation index: a cumulative list of articles that, subsequent to the appearance of an original article, refer to that article.

5. *Uniformity and compatibility are desirable (p. 36).*

Since the entire information system is a network of separate subsystems, rapid and efficient switching between the different elements of the system is essential. Such switching will be fully effective only if the different subsystems adopt uniform practices toward abstracting and indexing. We commend the Office of Science Information Service (OSIS) of the National Science Foundation for trying, through the National Federation of Science Abstracting and Indexing Services, to encourage order in a chaos of nonuniformity. We believe that Government, by virtue of the financial support it gives to private information services, should exert leverage in persuading societies to adopt more uniform practices.

B. *Recommendations to Government Agencies*

We preface our recommendations to the Federal agencies with the statement that Government information activities must not be allowed to swamp non-Government activities. The special sensitivity of non-Government, decentralized information services to the needs of the user as well as the variety of approaches offered by these services is precious and must be preserved. Support by Government does not necessarily mean domination by Government but this danger must always be guarded against.

1. *Each Federal agency concerned with science and technology must accept its responsibility for information activities in fields that are relevant to its mission. Each agency must devote an appreciable fraction of its talent and other resources to support of information activities (pp. 44ff).*

Since the information process is part of the research and development process, agencies that support research and development in fields that are relevant to their missions accept responsibility for supporting and otherwise carrying out information activities in these fields. Each of the mission-oriented agencies ought to become "delegated agents" for information in fields that lie within their missions. In these fields the agencies should maintain a strong internal information system and should support non-Government information activities, always striving to blend the Government and non-Government systems into a consistent whole.

2. *To carry out these broad responsibilities each agency should establish a highly placed focal point of responsibility for information activities that is part of the research and development arm, not of some administrative arm, of the agency (p. 45).*

We stress that the technical information activities of an agency must be part of research and development, not part of administration.

3. *The entire network of Government information systems should be kept under surveillance by the Federal Council for Science and Technology (p. 46).*

We applaud the recent action of the FCST in establishing an interagency Committee on Science Information. Among other matters, this committee will be expected to prevent overlaps and omissions as the agencies become delegated agents in various fields of science and technology.

4. *The various Government and non-Government systems must be articulated by means of the following information clearinghouses:*

   a. Current Efforts Clearinghouse (pp. 46–47). We recommend that the Science Information Exchange (that provides information on who does what where) be strengthened and that it receive separate support rather than depending on voluntary contributions from the agencies it serves. A Technological Efforts Exchange, either as part of SIE or working in close collaboration with it, should be established.

   b. Report Announcement and Distribution (p. 47). We recommend that the Office of Technical Services of the Department of Commerce be made a complete technical reports sales agency. It should be given enough support so that it can announce promptly and supply inexpensively a copy of any declassified Government technical report.

   c. Retrospective Search and Referral Service (pp. 47–48). We approve the recent action of NSF and the Library of Congress establishing a National Technical Referral Center as part of the Library of Congress.

   In addition, the National Referral Service should maintain and make available a directory of Specialized Information Centers and a register of formal technical meetings.

5. *Each agency must maintain its internal system in effective working order (pp. 38–43).*

The internal communication system is based largely on informal technical reports. We offer the following recommendations for improving the dissemination and retrieval of information contained in the technical reports:

   a. Technical reports should be refereed or otherwise screened before they enter the internal information system (pp. 39–40).

   b. Agencies must insist that their contractors live up to their contractural obligations for adequate technical reporting. We believe that proprietary interests sometimes serve as barriers to proper flow of information. We recommend that the whole matter of defining what are and what are not proprietary rights in Government contracting be subjected to a Government-wide study (pp. 41–42).

   c. Although the Panel sees no cause for alarm in the way classification is now handled by Government agencies, this impression is largely an intuitive one. We therefore recommend that problems of security and declassification be studied by an ad hoc group of the Federal Council's Committee on Information (pp. 41–42).

d. Since the report literature is often poor, critical reviews of the report and related literature play an important role. Critical review journals published under Atomic Energy Commission auspices have been generally successful; we urge other agencies, notably National Aeronautics and Space Administration and Department of Defense, to undertake similar review ventures in fields of interest. Such review journals might well become a most important product of the specialized information centers.

e. We believe that the large central agency depository should concentrate on being a document wholesaler, and that, where specialized centers exist, the job of preparing state-of-the-art reviews, and otherwise interpreting the literature, should be the responsibility of the specialized information center (pp. 43–44).

f. Since these latter activities are so important to the effective transfer of information, we believe that the agencies concerned should actively sponsor and support additional specialized information centers at appropriate establishments (pp. 33, 43).

6. *Problems of scientific information should be given continued attention by the President's Science Advisory Committee (p. 51).*

The problems of scientific information are very complex and they will continue to be with us. We therefore recommend that scientific information, and particularly the balance between Government and private activities, be given continued attention by the President's Science Advisory Committee.

PRESIDENT'S SCIENCE ADVISORY COMMITTEE

John Bardeen, Professor of Electrical Engineering and Physics, University of Illinois

Harvey Brooks, Dean, Division of Engineering and Applied Physics, Harvard University

Paul M. Doty, Professor of Chemistry, Harvard University

Richard L. Garwin, Watson Research Laboratory, Columbia University—International Business Machines

Edwin R. Gilliland, Professor of Chemical Engineering, Massachusetts Institute of Technology

Donald F. Hornig, Professor of Chemistry, Princeton University

George B. Kistiakowsky, Professor of Chemistry, Harvard University

Robert F. Loeb, Bard Professor of Medicine, Columbia University

Colin M. MacLeod, School of Medicine, New York University

Wolfgang K. H. Panofsky, Director, Stanford Linear Accelerator Center, Stanford University

Frank Press, Director, Seismological Laboratory, California Institute of Technology

Edward M. Purcell, Professor of Physics, Harvard University

Frederick Seitz, President, National Academy of Sciences

John W. Tukey, Professor of Mathematics, Princeton University

Alvin M. Weinberg, Director, Oak Ridge National Laboratory

Jerrold R. Zacharias, Professor of Physics, Massachusetts Institute of Technology

Jerome B. Wiesner, Special Assistant to the President for Science and Technology, The White House (*Chairman*)

Appendix F: Excerpts from the SATCOM Report

See National Academy of Science, Committee on Scientific and Technical Communication (SATCOM), *Scientific and Technical Communication: A Pressing National Problem and Recommendations for Its Solution, a Synopsis* (Washington, D. C., National Academy of Science, 1969), pp. 30.

Specific Courses of Action: SATCOM'S Recommendations

To implement the objectives outlined in the preceding section, we have developed 55 recommendations that deal principally with the management, performance, and economics of the nation's diverse but interrelated scientific-and-technical-communication efforts. Although we have tried wherever possible to place responsibility for needed action on particular organizations or agencies, we did not subject the activities of any specific group to individual criticism. We urge those engaged in developing and operating information services and in marketing their products to review the recommendations in their entirety, not only in this brief and nontechnical summary, but in Chapter 3 of our final report, in which they are presented in full with accompanying discussion.

The five general areas treated in our 55 recommendations include:

1. Planning, coordination, and leadership at the national level
2. Consolidation and reprocessing—services for the user
3. Classical services (abstracting and indexing, libraries, formal and semiformal publication, and meetings)
4. Personal informal communication
5. Studies, research, and experiments

PLANNING, COORDINATION, AND LEADERSHIP AT THE NATIONAL LEVEL

Because we believe, as we indicated at the end of the preceding section, that a broadly representative, nongovernmental body of high prestige is essential to stimulate greater coordination among private groups and to facilitate their interaction with appropriate branches of the government, the first of our recommendations calls for the establishment of a *Joint Commission on Scientific and Technical Communication, responsible to the Councils of the two Academies.*

Also mentioned in the preceding section was the basic philosophy of *shared responsibility between governmental and private organizations for the effective communication of scientific and technical information.* This is the subject of our second major recommendation, which calls on the organizations of the scientific and technical communities to recognize the national implications of their activities and the proper role of the government and, in turn, calls on government agencies to allot a central place in the management of the information services required in support of agency missions to the scientific and technical community or in some cases to commercial information-handling organizations. An important application of this principle is found in information programs that, though needed for the accomplishment of a government-agency mission, are directed in major part to workers outside the government (i.e., information programs that are principally discipline-oriented). We advocate that *such government-sponsored but discipline-oriented programs be managed by appropriate scientific and technical societies, by federations of such societies, or, in special cases, by commercial organizations.* As an illustration, we mention the National Aeronautics and Space Administration (NASA), which fostered the creation of the American Institute of Aeronautics and Astronautics (AIAA), a nongovernmental organization of scientists and engineers. Operating under contract to NASA, the AIAA makes information on journals, books, and meeting publications in this field available through *International Aerospace Abstracts*. In addition, coverage of the worldwide technical report literature is provided through *Scientific and Technical Aerospace Reports,* which is managed by a commercial organization under contract to NASA.

A particularly important corollary of the philosophy we are enunciating is that those who support research-and-development work have a responsibility to see to it that the information so generated becomes truly available. Because it was realized some years ago that work does not properly benefit society until it is published and accessible, federal research grants now provide for support of an appropriate share of the costs of publication. But today mere publication of isolated tidbits scattered through a multitude of journals is rarely sufficient to place the totality of new results and insights on a given subject effectively within the grasp of those who should benefit from them. So we now call for acceptance of a broader responsibility and urge *sponsors of research-and-development work to recognize as integral to the support of such work the publication and processing of the information so generated for access, consolidation, and use in special contexts.*

Although to be practical our recommendations have to be addressed almost entirely to governmental and private organizations in the United States, we must never lose sight of the fact that scientific and technical communication is a world activity, not just a national one. The contribution made by the United States to the world's primary literature has always been only a fraction of the total, and this fraction is decreasing as more countries achieve high productivity. Thus, the need for international cooperation is already great and will continue to grow. Leadership in our so-called national programs will increasingly involve the development of more-effective international scientific and technical communication; therefore, we recommend that *the policy-making groups of our scientific and technical societies encourage the managers of their major information services in the development of ways in which access and transfer activities can operate on a more truly international basis through sharing both work and products across national boundaries.* Related recommendations point to (a) the responsibility of the federal government to encourage and assist private efforts to effect international information-exchange arrangements; (b) the need for early planning by the management of international research efforts (such as the International Biological Program) for the storage and transfer of the information and data so generated; and (c) the necessity of including representatives of relevant nongovernmental information activities in U.S. delegations to internationally managed information projects.

Two final recommendations on national planning deal with special problem areas of widespread concern—copyright and standardization. Both are complex and include a wide range of problems other than those related to scientific and technical communication. Each demands careful study over its whole range. In the case of copyright, recent congressional hearings have underscored the fact that existing copyright law is not adequate to cope with the problems posed by the radical new techniques now available for reproducing documents and for storing and processing information. We believe that this field should be studied in depth before necessary new legislation is developed to deal with these problems. Therefore, we support the concept of legislative action to create a special statutory commission to study copyright problems (entirely distinct, of course, from the Joint Commission proposed in our first recommendation).

Standardization has to do with how a file of information generated by one group can be efficiently accessible to another, especially if communication between electronic computers is involved. As vast computer-based stores are already accumulating rapidly, no time should be lost in working out and securing general acceptance of carefully considered standards that will minimize technological difficulties in this kind of communication. Therefore, we advocate the creation of a special task group of the proposed Joint Commission on Scientific and Technical Communication (of the two Academies) to maintain awareness of relevant developments in standardization and convertibility.

CONSOLIDATION AND REPROCESSING— SERVICES FOR THE USER

Often it is of very little help to a worker, be he scientist, engineer, or physician, merely to supply him with a long list of publications relevant to the problem with which he is grappling. What he needs is something that will organize and evaluate what is known about a subject and present it in language that he can understand and at the level of detail that he wants. Such consolidations of information, the preparation of which often requires great intellectual creativity, have traditionally appeared in review articles, books, data compilations, and the like. But the preparation and use of such material has not kept pace with the flood of potentially useful new information in the scientific and technical literature. So we not only call for acceptance of broader responsibilities in this area by the sponsors of research and development, as indicated in the preceding group of recommendations, but we also urge *scientific and technical societies to take major responsibility for identifying needs for critical reviews and data compilations, furthering their preparation, fostering awareness of their existence, and stimulating education in their use.* In addition, so great are needs in this area that the proposed Joint Commission should take a hand in stimulating consolidation efforts and suggesting feasible ways of promoting easier and more-efficient access to those reviews appropriate to a user's specific needs.

Another crucial area is that of providing the practitioner—the person who has the job of putting new knowledge to practical use in industry, hospitals, farms, and the like—with the types of information that he requires, presented in the language and format that will be most meaningful to him. Currently, much is written about the lag between discovery and application—between the announcement of new knowledge and its incorporation in new technological developments. The provision of specialized services geared to the needs of the practitioner, and to subfields, and cross-disciplinary areas of need, are key steps in speeding the effective use of information. We recommend that *societies whose membership includes large numbers of practitioners, especially in engineering, medicine, and agriculture, increase their attention to information programs that will ensure awareness of and access to information of particular interest to practitioners; identify and stimulate efforts to fulfill the needs for state-of-the-art surveys, reviews, and data banks; and provide for the practitioner's need for continuing education to keep up to date in special fields.* Since many of these types of services traditionally have been handled by for-profit organizations, scientific and technical societies should encourage such organizations to undertake them.

Several more of our recommendations deal with ways in which the design and initiation of special information services for diverse need groups (as defined in earlier sections) can be facilitated by scientific and technical societies, by the government, and by the proposed Joint Commission. Once such services are established, they should usually be able to gain their support from the users they serve. Private enterprise and small-group interests can be very effective in discerning and meeting the needs of diverse groups, but only if the information that has to be reprocessed for the necessary services can be made available to them on economically reasonable terms. Much of the information will usually come from the broader discipline-oriented basic abstracting and indexing services, which means that, ideally, the products of the latter services should be available to all who would like to use them at costs reflecting merely the extra effort necessary to supply the services in quantity to the new customers. Therefore, it is urgent that *those societies and agencies concerned with the conduct and support of abstracting services should seek actively to identify difficulties, find solutions, and take the initiative in proposing and testing arrangements through which an increasing contribution by the sponsors of research to the input costs of the basic abstracting services can make transfer for reprocessing financially feasible at approximately output (reproduction and distribution) costs.* The accomplishment of this objective will be neither easy nor rapid. In the interim, basic abstracting and indexing services must be responsible for launching reprocessing efforts or stimulating the effective use of their products. And scientific and technical societies must recognize and prepare to assume their responsibilities for adequate reprocessing of access information in their respective disciplines. The aid of commercial organizations should be actively sought for the fullest development of useful reprocessing services.

Even as access to reviews presents a problem, so does access to the basic abstracting and indexing services appropriate to the particular needs of diverse user groups; therefore, one of our recommendations is addressed to the exploration of means of guiding users in their choice of the indexing and abstracting services appropriate to their needs.

A considerable potential for the provision of the specialized need-group services we have described exists in the form of information analysis centers. These centers, of which over 100 are now sponsored by the federal government, have been set up to serve certain specific fields in which there are large amounts of data that require critical evaluation. Such a center typically consists of specialists who collect, assemble, evaluate, and store information about a certain subject area and make it available to specific groups of users. We feel that the capabilities of these centers are not fully utilized, and we recommend that the proposed Joint Commission aid in identifying the major information analysis centers dealing with particular subject areas and capable of offering services to specialized need groups. It also should stimulate the exploration of ways in which such services can be made more widely available.

Two final recommendations in this section treat management problems, which, although they are encountered in all forms and on all levels of scientific-and-technical-communication activities, are especially prominent in providing highly user-oriented services. The development of adequate and continuing feedback mechanisms to assure the relevance of services and the need for increased emphasis on substantial marketing and educational efforts to overcome the "in a rut" and "line of least resistance" behavior patterns of users faced with new and improved services are highlighted in these recommendations.

THE CLASSICAL SERVICES

We include among the classical services:

1. Basic abstracting and indexing
2. Selection, acquisition, bibliographic control, reference, housing, document delivery, and other service functions of libraries
3. Formal and semiformal publication
4. Meetings

Most areas of science and technology have developed the custom of keeping track of the literature of their fields by publishing comprehensive collections of abstracts—usually one-paragraph summaries of articles published in specialized journals—and by suitably indexing this material. The increasing volume of material to be covered has caused the cost of such services to rise rapidly. At the same time, increasing subscription costs have discouraged individuals (as opposed to libraries and other institutions) from subscribing to them.

In seeking new ways to support abstracting and indexing services and to improve existing support mechanisms, special care is necessary to ensure the broad usefulness of the product and maximum responsiveness to the progress of science and technology. Two paramount issues in this context are sensitivity of management, particularly in regard to scope of coverage and adequacy of abstracting and indexing, and availability of abstracts for reprocessing.

In addition to the basic abstracting and indexing services we have described, which systematically order for permanent reference all material published in a given discipline, numerous other services exist that are also extremely useful. Examples are listings of titles of items newly published and citation indexes, which show the interrelationship among a wide selection of items by indicating which ones cite others. These types of services differ in nature and scope from basic abstracting and indexing and need not be managed in the same way.

Our central recommendation with regard to these so-called secondary services undertakes to preserve a proper balance between the interests of the nation, and especially the federal government, in ensuring uninterrupted availability of the basic services and the advantages of leaving the management of both kinds of services in the hands of those who will be most sensitive to the needs of scientific and technical disciplines or appropriate groups of users. Our recommendation is that the *departments and agencies of the federal government fund the literature-access services that are needed for the effective utilization of the knowledge resulting from the research and technical activities that they sponsor. In so doing they should ensure (a) management of basic discipline-wide abstracting and indexing by appropriate scientific and technical societies or federations thereof, though the use of for-profit services in special cases should not be precluded; and (b) management of other broad bibliographic services (e.g., title listings) by private for-profit organizations, national libraries, or societies.*

There follow two recommendations on steps that scientific and technical societies, aided by the National Federation of Science Abstracting and Indexing Services, can take to improve the quality, timeliness, and efficiency of preparation of the necessary abstracts and indexes.

The role of libraries in the dissemination of information and in evolving national information systems was the focus of the recent intensive survey of the National Advisory Commission on Libraries. The basic import of the five recommendations presented in their report is in harmony with SATCOM's over-all approach and recommendations, two of which deal with libraries. We feel that it is imperative that library services be made much more responsive, and that there are few limits to what can be done if adequate resources are made available. However, pouring more and more money indiscriminately into libraries will not solve their complex problems. Therefore, we recommend a funding policy aimed at introducing: (a) a more-realistic reflection of library costs in the conduct of scientific and technological work; (b) a closer relationship between costs and services; and (c) more options of extra service for an extra price. We advocate *direct grants from appropriate agencies for the strengthening of research-library services, with emphasis on funding new and innovative services and special provisions in research grants to educational and research institutions for adequate funds for the use of needed library and information services. These funds should be provided in such a way that researchers can exercise discretion in their use and choose the services that they find most valuable.* Our second recommendation with regard to libraries focuses on education—the education of users in the existing array of library services, and training programs that place greater emphasis on the operational analysis of library services.

Our next recommendations deal with formal primary publications, that is, the scientific and technical journals in which new results and discoveries are presented to the professional public. The relentless expansion in the amount of new material to be published has posed intellectual problems for the journals and has involved them in growing financial difficulties. There is pressing need for systematic study and analysis of the economics of these publications. Consequently, we recommend *studies to examine the income returned to such publications from their principal markets—users, authors, and the public—together with trends in cost factors and the impact of new technologies to guide the evolution of flexible funding and pricing policies that will be responsive to the needs of each interested party without being unduly responsive to any.* Until such data have been assembled and possible alternative funding arrangements have been convincingly tested, we urge that *sponsors of research and development continue to provide support through page charges* (see discussion in the section on "Roles and Responsibilities") *for the publication of such work.*

Several subsequent recommendations offer suggestions for reducing publication delays, using new advances in the techniques of publication, directing highly selective material to individuals or small groups, and disseminating information on "who is doing what, and where."

In addition, two recommendations deal with semiformal communication (i.e., technical reports or papers prepared for publication and circulated before their appearance in journals). There has been much controversy over the purposes of such modes of communication. Many people argue that everyone interested should have access to any information that can be supplied without having to wait for formal publication. On the other side are those who fear undermining of the tradition of placing new results in journals so that they can be identified and located for ready reference and where certain standards of quality and novelty must be met. We believe that there is merit in both arrangements and that these views are not entirely irreconcilable. Therefore, we urge (a) adequate bibliographic control, insofar as is practicable, of semiformal publications that need wide distribution so that they can be readily identified; (b) general accessibility through storage in depositories; and (c) control of distribution to the extent necessary to protect formal publications. We also advocate a clearer differentiation between those technical reports required at certain intervals, regardless of the status of work (e.g., periodic progress reports), and substantive reports prepared when the work warrants. Additionally, standards of uniformity in the documentation of all substantive reports are needed to allow adequate bibliographic control and to foster accessibility.

PERSONAL INFORMAL COMMUNICATION

In a qualitative sense we know that personal informal exchanges play a major role in the transfer of scientific and technical information. The increasing number of informal information-exchange groups, the steadily growing tendency toward collaborative and team research, and the current emphasis on conferences, meetings, interinstitutional visits, and other occasions that facilitate informal interaction are evidence of both general awareness of the necessary role of informal communication and increasing dependence on it. Such emphasis and dependence have resulted in numerous studies to determine the characteristics, content, and functions of informal interpersonal communication. But accurate, quantitative comparison of the effectiveness of the very informal interpersonal techniques of communication with that of other communication methods and media is difficult to achieve and presents many problems. Though progress is under way in this area, far more clear-cut and comprehensive data on the ways in which informal channels operate are necessary before constructive recommendations to enhance its effectiveness can be developed. Therefore, we offer but two recommendations on this topic; one stresses the importance of providing ample opportunities and facilities for informal communication at scientific and technical meetings, and the other encourages leave and sabbatical policies that foster interinstitutional visits and exchanges of personnel.

STUDIES, RESEARCH, AND EXPERIMENTS

In a final group of ten recommendations, we suggest some studies and experiments that are urgently needed and some guidelines for their conduct. We give *top priority to the initiation of comprehensive analyses and of experiments on the functioning of different parts of the scientific-and-technical-communication network and on its over-all operation.* Efforts to develop *measures of the value of information services and ways of over-*

coming user apathy or resistance in the face of new options and services also receive major emphasis. Comparisons of various means of storage and transmission and the careful consideration of their implications for information-handling practices—for example, the question of centralized versus decentralized depositories—deserve special attention.

As we have already indicated in several places in the preceding paragraphs, a wide range of exciting possibilities for the use of new technologies in the storage, processing, and transmission of information exists. Though it may be Utopian to dream of meeting all needs by simply "asking the computer," there can be no question that the scientists and technologists of the future will be able to handle information in ways that were undreamed of because of the rapidly increasing capacity of computers to store and search vast amounts of material. Carefully planned applications and modifications based on experience will be necessary if the full potential of new technologies is to be realized. Therefore, we urge a series of major experiments involving the use of the computer in conjunction with human workers for the preparation of indexes; "evolutionary indexing" in which a small widely used file of references on some single subject area can have basic critical comments added to it by qualified users—comments that can benefit subsequent users; the development and evaluation of languages for describing the formats of files and other types of digital communication systems; and the development of standard structures for each widely used bibliographic documentary information element. The participation of highly competent scientists, engineers, and practitioners is of vital importance to ensure the relevance of such experiments to key questions and issues, coherence of effort in their conduct, and efforts to follow up and apply their findings. *The responsibility of the scientific and technical societies to encourage such participation receives special emphasis.*

Finally, large-scale experiments that involve large populations and the use of advanced technologies are necessary for the fulfillment of increasingly diverse needs; such experiments constitute exploratory development as well as research, and they require special provisions for planning and funding. We recommend that the *federal government establish a single group to (a) plan a unified program of critical experiments of operational scale in scientific and technical communication and (b) find, guide, and support contractors in the conduct of such experiments.*

The studies, research, and experiments that we have explicitly advocated are directed only to the most urgent needs; a continuing flow of work on a wide variety of problems is essential to progress. Our recommendations are intended merely to serve as points of departure and to challenge increased attention and effort.

Appendix F

Committee on Scientific and Technical Communication (SATCOM)

Robert W. Cairns, *Chairman*
Vice President
Hercules, Inc., Wilmington, Delaware

Jordan J. Baruch (from February 1968)
President
Interuniversity Communications
 Council (EDUCOM)
Boston, Massachusetts

Curtis G. Benjamin
 (November 1966–April 1968)
Special Consultant
McGraw-Hill Book Company
New York, New York

Raymond L. Bisplinghoff
 (from July 1966)
Dean of Engineering
Massachusetts Institute of Technology
Cambridge, Massachusetts

Daniel I. Cooper
 (May 1967–March 1968)
Publisher (formerly)
International Science and Technology
New York, New York

Ralph L. Engle, Jr., M.D.
 (from April 1967)
Associate Professor of Medicine
Cornell University Medical College
New York, New York

Conyers Herring
 (from December 1966)
Research Physicist
Bell Telephone Laboratories
Murray Hill, New Jersey

George E. Holbrook
Vice President
E. I. du Pont de Nemours & Co., Inc.
Wilmington, Delaware

Donald L. Katz (from February 1968)
Department of Chemical and
 Metallurgical Engineering
The University of Michigan
Ann Arbor, Michigan

J. C. R. Licklider
Director
Project MAC
Massachusetts Institute of Technology
Cambridge, Massachusetts

Clarence H. Linder
Vice President and Group Executive
 (retired)
General Electric Company
Syracuse, New York

Jerome D. Luntz (from April 1968)
Vice President for Planning and Development
McGraw-Hill Publications
New York, New York

H. W. Magoun (through May 1967)
Dean, Graduate Division, and Professor of Physiology
University of California at Los Angeles
Los Angeles, California

Oscar T. Marzke
(August 1966–February 1968)
Vice President
Fundamental Research
United States Steel Corporation
Pittsburgh, Pennsylvania

Nathan M. Newmark
Head
Department of Civil Engineering
University of Illinois
Urbana, Illinois

William H. Pickering
(through June 1966)
Director
Jet Propulsion Laboratory
California Institute of Technology
Pasadena, California

Byron Riegel
Director of Chemical Research
G. D. Searle and Company
Chicago, Illinois

William C. Steere
Director
New York Botanical Garden
Bronx, New York

Don R. Swanson
(from November 1966)
Dean
Graduate Library School
University of Chicago
Chicago, Illinois

John W. Tukey
Professor of Mathematics
Princeton University
Princeton, New Jersey

Merle A. Tuve
Director
Department of Terrestrial Magnetism
Carnegie Institution of Washington
Washington, D.C.

Paul Weiss (through May 1966)
University Professor
Graduate School of Biomedical Sciences
University of Texas
Austin, Texas

W. Bradford Wiley
(through October 1968)
President
John Wiley & Sons, Inc.
New York, New York

Irving S. Wright
(August 1966–March 1967)
Professor Emeritus, Clinical Medicine
Cornell University Medical College
New York, New York

Van Zandt Williams
(through May 1966) (deceased)
Director
American Institute of Physics
New York, New York

STAFF OF THE COMMITTEE ON SCIENTIFIC AND
TECHNICAL COMMUNICATION (SATCOM)

F. Joachim Weyl, *Executive Secretary*
Bertita E. Compton, *Staff Officer*
Judith A. Werdel, *Staff Associate*

Appendix G: Excerpts from the Kennedy Report

SUMMARY OF FINDINGS

The strength and viability of our Nation's scientific and technical information services—both within the Federal Government and the private sector—have attracted the attention of the Congress from time to time. Since the late fifties, Members and committees of Congress have taken action resulting in measures creating certain information programs, performed traditional oversight by examining department and agency programs, and held hearings designed to focus on salient information handling problems and practices.

Yet the importance of scientific and technical information as a national resource has only recently emerged. It is difficult for nontechnical personnel to understand that one of the most significant products of the sizable Federally-funded research and development programs is information. These voluminous data and accompanying narrative insights, presented in varying forms, can be fully utilized only through a strategy for their generation, management, and disposition. This strategy is not apparent in today's STINFO operation. To fulfill these interlocking tasks implies the presence of an assertive leadership, aware of the complex needs of this post-industrial, civilian-oriented society and dedicated to ensuring that that combination of authority, responsibility, and flexibility necessary to carry out this priority mission is functioning. Today, there is a widely alleged lack of such leadership, suggesting strongly that "management" of a precious resource is fragmentary and questionable at best.

What should be the nature of such leadership and how should it be served by advisory mechanisms, representing both the "offerors," "users," and "managers" of information? Recommendations have been

---

See U. S. Congress, Senate, Special Subcommittee on the National Science Foundation of the Committee on Labor and Public Welfare, Federal Management of Scientific and Technical Information (STINFO): The Role of the National Science Foundation, report prepared for the Special Subcommittee by the Congressional Research Service of the Library of Congress, Committee Print, 94th Cong., 1st sess. (Washington, D. C.: Government Printing Office, July 1975), pp. 103.

forthcoming from commissioned studies and gatherings of the leaders of the community concerning the desirability of creating a strong coordinating mechanism, which would involve resources of both scientific and non-scientific, public and private groups. The consensus was that such a body must be prepared to implement policies which serve the needs of Government, industrial, and university STINFO activities. More specifically, emphasis would be placed on optimizing cooperation and minimizing duplication in such areas as:

- creating the type of linkages (e.g., on-line networks) which improve cooperation between information centers, traditional libraries, those who publish, and the varied user groups.

- operation of indexing and abstracting services, the issuance of primary journals and more innovative dissemination mechanisms.

- reducing overlap between and maximizing capabilities of services offered both by Federal information systems and their private sector (both profit and non-profit) counterparts.

Many agencies of the Federal Government, it is alleged, have tended to become preoccupied with their own programs and resist cooperative endeavors--"clinging to prerogatives" was the term employed by one Congressman--resulting in a feudal posture which impeded the realization of national goals. The private sector groups, whether "wholesalers" or "retailers" of information, understandably want to have a say in determining their own future, or to put it another way, "They don't want to be surprised."

With the disappearance of COSATI, OST, and the Science Information Council, management mechanisms are few. As noted above, there is an often articulated need for a new organization capable of performing deliberative, analytical tasks as well as simply monitoring and coordinating. In other words, more than a forum for information exchange was visualized. The composition of this advisory organization should be broadly representative of the key elements making up the STINFO community, but not hinged upon specific organizations or specialties: professional society representatives, managers of STINFO activities and R&D "laboratories," information/telecommunications scientists, scientists and engineers in the roles of producers and consumers, managers from information industries, and that genre comprised of policy and program analysts. With this representative mix, the benefits to be derived from a formal advisory mechanism could result in decision-making more attuned to the real world and the opportunity for meeting interdisciplinary, social needs would seem to be greater.

The options related to this projected coordinating body have been enumerated at various times. A filtering of those would bring forth this kind of composite recommendation:

- an advisory group reporting directly to the President's Science Adviser in his role as Director of NSF, thus being able to meet national responsibilities under Title IX of the National Defense Education Act.

- a membership of senior, prestigious personnel, numbering less than a dozen, supported by a high quality professional staff.

- creation of this group in the immediate future, possibly with a legislated mandate, with the potential of imbedding it in any future reconstituted S & T office connected to the White House.

With this type of ingrained expertise and secretariat, the advisory group not only could review the past and monitor the present, but provide guidance responsive to projected needs and goals. Its arena would include providing factual and interpretive contributions to those engaged in high-level policymaking and such "shirt-sleeves" functions as determining information exchange criteria (both inter-system and international), identifying "deliverables" from existing systems, hammering out long-needed standards, calculating long-range user needs, and employing innovative information technologies. To iterate a point made earlier, each action area must be viewed as a "tie-point" to reality, with no time for academic niceties.

In highlighting the priority functions of this advisory group, the past must serve as prologue in attempting to avoid pitfalls--for example, responsibility for providing coordinative leadership, but no real authority--and capitalize upon already existing, thoughtful recommendations. Judging from a study of special reports, proceedings from now-defunct advisory group meetings, and records of interview with key community personnel, a series of essential duties are suggested:

1. Offering advice, based on analysis and deliberation, suitable for shaping policy, executing program review and assessment, and budgeting preparation;

2. Developing lasting but flexible monitoring and coordinating procedures, including a "feedback" capability to ensure responsiveness;

3. Evaluating, in detailed fashion, existing STINFO services and products, particularly in terms of their quality and broad availability and publishing these findings;

4. Periodically looking at Federal Government and private sector information services with an eye to user needs, expressed in terms of quality of product, cost, and competitive stance;

5. Working to ensure a reasonable balance in Federal funding of Government and non-Government R&D in information services;

6. Performing and contracting research and development on scientific and technical information products, services, and systems; and

7. Providing an interface with private and international STINFO activities.

Assuming the establishment of an action-oriented advisory group, then the mission and program objectives of the NSF Office of Science Information Service could merit re-examination.

Budget levels for scientific and technical information services have often been tied in theory--but never in fact--to overall R&D expenditures. Various witnesses and commentators have offered figures ranging from three to ten percent as a "proper balance." As noted, however, overall agency STINFO figures (using the NSF survey) account for less than three percent of Federal R&D expenditures. The annual decrease in OSIS budget allotments over the last few years also raises questions regarding that Office's capabilities to exercise any planning or coordination functions in the Federal or non-Government environment for scientific and technical information.

Because of newly defined national goals and objectives, Government and society have new requirements for information relevant to STINFO services. There is an increasing urgency for the Congress, OMB, the Federal STINFO community, and the diverse university-society-industry triad to have a focus for shaping their programs, R&D requests, budgets, and for maximizing the use of existing and future information resources. In response to this need for information and the relatively fragmented structure of existing STINFO systems, a series of questions have emerged during this study for consideration by the Special Subcommittee:

- Is there a requirement today for Government oversight of private sector information services, including those originally subsidized by Federal funds?

- Would it be valuable to obtain testimony from key STINFO figures--both at the policymaking and operational levels--in order to update the overall perspective of the Government management role, which might be helpful in modifying or reinforcing past recommendations for improvement of service?

- Does the National Science Foundation view its management and coordination roles as achievable in the light of the sharp decrease (in the 1970 decade) in its OSIS personnel and funding support?

- Could a balanced view of the "health" of (both public and private) STINFO services and responsibilities in terms of developing optimal "systems" responsive to national goals and international commitments be attained by convening a congressionally-sponsored colloquium?

- With Federal budgetary pressures demanding a close scrutiny of all Governmental expenditures, would an assessment of the service provided for full range of user groups by existing information services be helpful in developing guidelines which might eventually result in a better "mix" of such public and private sector services and systems?

- Inasmuch as there is a continuing need to alert and educate key executive and legislative leadership personnel to trends-over-time in STINFO requirements and services, would it be useful to have NSF release for general distribution those recent studies thus far restricted?

Index

Abstracting Board of the International Council of Scientific Unions (ICSU-AB), 128
Abstracting journals, 189
Abstracting services, 72, 192
  access to, 212
  British, 142
  Heller report on, 104-105
  of NSF, 104
Adams, Scott, 155
Addiction Information Center, 81
Adkinson, Burton W., 57, 156
Administrators
  R&D, 30, 94, 144
  in sci-tech information, 155-161
  of Smithsonian, 16
Advance Research Projects Agency (ARPA), 74, 88, 90
Advisory Committee for Scientific and Technical Information (ACSTI), British, 141
Advisory panels, nonfederal, 118-119
Aeronomy and Space Data Center, 84
Agency for International Development (AID), 36, 132, 175
Agriculture, Department of. *See also* National Agricultural Library
  developments in, 24-26
  information of activities, 65
  international programs of, 133
  libraries of, 10, 12, 49-50, 160, 171
  "plant explorers" of, 131
  University of California, 116
Agriculture Research Service, 37
Agri-Doc, 7
Aines, Andrew A., 156
Air Documents Division, 150. *See also* Central Air Documents Office
Air Documents Research Center, 31, 173
Air Force, Office of Scientific Research, 74, 88, 90

Air Pollution Technical Information Center, 81
Alexander, Samuel, 156
Allen, Ernest H., 40
All-Union Institute of Scientific and Technical Information (VINITI), 2, 53, 137
American Association for the Advancement of Science (AAAS), 113, 126
American Association of Geologists (AAG), 113
American Chemical Society (ACS), 113, 124, 149
American Documentation Institute, 26, 128
American Geographical Society (AGS), 113
American Geological Institute, 123, 124
American Institute of Aeronautics and Astronautics (AIAA), 116, 209
American Institute of Electrical Engineers (AIEE), 113-114
American Institute of Physics, 124
American Institute for Research, 85
American Medical Association, 116
American Physical Society (APS), 113
American Psychological Association, 124
American Society for Metals (ASM), 46, 143
Ames Laboratory, 178
Argonne National Laboratory, 115
Armed Services Medical Library, 49, 149, 160, 173
Armed Services Technical Information Agency (ASTIA), 32-33, 41, 61, 150, 159, 161, 175
  computer use of, 70
  copyflow units for, 119
Army, early expeditions of, 23
Army Medical Department, 23
Army Research Office, 88
Army Signal Corps, 10, 19
Arnhym, Albert A., 31
Astin, A. V., 99

225

Atomic Energy Act, 43-44, 48
Atomic Energy Commission (AEC)
   cooperative exchange projects of, 48, 132
   expansion of, 30
   R&D facilities of, 115
   technical information center in, 61
   Technical Information Division, 5, 42-48
   Technical Information Panel, 118
Auerbach Corporation, 121
Authors, responsibility of, 201
Automated systems, introduction of, 86. *See also* Technology
Automatic data processing (ADP) standards, 163
Automatic Picture Transmission (APT), 136

Bache, Alexander Dallas, 21
Bacher, Robert F., 194, 195
Baird, Spencer Fullerton, 18, 19
Baker, William O., 54, 100, 194, 195
Baker Panel, 175
Baker Report, 54, 185-195
Bardeen, John, 206
Baruch, Jordan J., 218
Benjamin, Curtis G., 218
Berkner, Lloyd V., 194, 195
Bethe, Hans A., 194, 195
Bibliofilm, 173
Bibliographic systems, 153
   LOC influence on, 15
   on-line interactive, 80
   selective-dissemination, 153
   technological innovations for, 47
Bibliography of Agriculture, 49
Bibliography of Economic Geology, 126
"Bibliography of Scientific and Technical Reports," 34
Billings, John Shaw, 23, 172
Billings, J. W., 12
Biological Abstracts, 124, 126
Biological Sciences Information Exchange (BSIE), developments in, 39-40
BioSciences Information Exchange, 175
Bisplinghoff, Raymond L., 218
Boardman, Brewer F., 45
Brady, Edward L., 156
Brenner, E., 100
British Library Act, 142
British Royal Society, 1
Bronk, Detlev W., 194, 195
Brooks, Harvey, 206

Brown, Harrison, 107
Brown, Jack E., 140
Brunenkant, Edward J., 47, 58, 157
Brussels Treaty, 15
Burchinal, Lee G., 82, 157
Bureau of American Ethnology, 19
Bureau of Budget, 6
Bureau of Chemistry, DOA, 25
Bureau of Commercial Fisheries, 62
Bureau of Marine Fisheries, 115
Bureau of Mines
   expansion of, 30
   information collections of, 10
Bureau of Naval Electronics, 61
Bureau of Ships, 61
Bureau of Standards, 5, 163
   Computer Division of, 156
   expansion of, 30
   information collections of, 10
   sci-tech information services of, 62
Bush, Vannevar, 31, 96

CADO. *See* Central Air Documents Office
CAIN (Cataloging and Indexing) system, 66, 153, 178
Cairns, Robert W., 218
Canada
   National Research Council of, 1
   National Science Library of, 2, 105
   sci-tech information programs of, 140-141
CANCERLINE, 82
Cannegeiter, Henrik Gerrit, 135
Cannon, Keith, 37
CAN/SDI program, 141
Carlson, Walter M., 61, 157
Carmaichel, Leonard, 40
Carnegie Corporation, 125
Cataloging, technological innovations for, 47, 70
Central Air Documents Office (CADO), 32, 150, 151, 174
Central depositories, proposed, 202
Central Intelligence Agency (CIA), 74
Chemical Abstracts, 20, 81, 123, 149, 174
Chemical-Biological Coordination Center, 127
Clapp, Verner W., 157
Classification systems
   geological, 22
   LOC, 13
   and technological innovations, 66, 70

Index   227

Clearinghouse for Federal Scientific and Technical Information (CFSTI), 37, 58, 62, 84, 120, 158, 160, 178
Coast and Geodetic Survey, 5, 62, 162, 171
  developments in, 20-21
  expansion of, 30
Commerce Department, 3-4
  Office of Declassification and Technical Service, 34
  Office of Technical Services, 4, 34-35, 62, 132, 150
  sci-tech information services of, 62
  technical information center in, 61
Committee for Data for Science and Technology (CODATA), 7, 62, 117, 134-135
Committee for International Cooperation in Information Retrieval among Examining Patent Offices (ICIREPAT), 100, 143
Committee on Scientific Information (COSI), 57-59, 71, 176
Committee on Scientific and Technical Communications, NAS, 207-219
Committee on Scientific and Technical Information (COSATI), 56, 58-59, 85, 107, 156, 179
Committee on Standards, of American National Standards Institute, 124
Communication. See also Information sciences
  biomedical, 80
  personal informal, 216
Computerized information systems, 163
Computer Science and Engineering Board, Information Systems Panel of, 129
Computer systems, 75
  federal adoption of, 119
  introduction of, 121, 151
  SEAC, 97, 98
  UNIVAC, 121, 152
Congress, U.S.
  and advisory panels, 118
  and Library of Congress, 12
  and NAS, 125
  and Patent Office, 96
  and sci-tech information, 162
  and Smithsonian Institution, 9, 16
Congressional Research Service, 16
Consolidation, 162
Cooper, Daniel I., 218

Coordination, 162
Copyflow unit, 94, 119, 120, 152
Copying practices, royalty schemes for, 103-104
Copyright law, 13, 73
  new, 103
  problems of, 67, 210
Copyright Office, 5
Council of Mutual Economic Assistance (CMEA), 139
Council on Library Resources, 60
Crawford, James, 55
Crawford report, 55, 177, 197-198
Crawford task force, 57, 101-102
CRIST (Current Research in Science and Technology) system, 65-66
Cummings, Martin M., 80
Current Index to Journals in Education, 82
Current Research and Development in Scientific Documentation (CRDSD), 88, 95, 117

Data centers, 189
Data information analysis, 148
Davis, Ruth E., 80, 157
Day, Melvin S., 47, 65, 158
Dayton Air Documents Division, 31, 32
Defense Department
  Project Lex, 67, 71, 118
  technical information centers of, 61, 117
  technical information services for, 50
Defense Documentation Center (DDC), 5, 61, 84, 147, 157, 160
  developments in, 30-33
  Xerox equipment of, 120
Defense Topographic Center, 23
Deignan, Stella L., 38, 158
"Delegated Agency" concept, 102, 177
Department of Scientific and Industrial Research (DSIR), of United Kingdom, 2, 141
Depot of Charts and Instruments, 24
Documentation Incorporated, 29
Documentation, Indexing and Retrieval of Scientific Information, 176
Documents Service Center, ASTIA's, 32-33
Doolittle, James H., 194, 195
Doty, Paul M., 206

Economic Cooperation Administration (ECA), of State Department, 36
Economy, need for, 79

Education Resources Information Centers
   (ERIC) system, 82, 157, 178
Eisenhower, Dwight D., 48, 56
Electrostatic reproduction, 151
Elliot, Carl, 56
Elliot Committee, 177
Ellsworth, Henry L., 11
Ely, William J., 58, 59
Energy Resources Development Agency
   (ERDA), 162
Engineering Index, 124
Engineers, and federal R&D programs, 94
Engle, Ralph L., 218
Environmental Data Service, 83-84
Environmental Protection Agency (EPA), 90
Environmental Research Information
   Service, ESSA's, 90
Environmental Sciences Information Center,
   84
Environmental Sciences Service Agency
   (ESSA), 62, 178
"Era of Information Explosion," 75
European Atomic Energy Agency
   (EURATOM), 48, 133
European Space Research Organization
   (ESRO), 83, 134
Executive Order 10521, 54, 101
Expeditions, Naval, 24
Experts, short-term use of, 119

Federal Advisory Committee on Scientific
   Information (FACSI), 56-57, 71,
   101
Federal agencies. *See also specific agencies*
   and NAS, 7
   and private organizations, 164
   R&D programs of, 60-61
   sci-tech information programs of,
      114-118
   systems programs initiated by, 121
Federal Council for Science and Technology
   (FCST), 55, 107, 204
Federal departments, R&D programs of,
   60-61
Federal government
   and nongovernmental organizations,
      136-137
   and support of nonfederal institutions,
      113-114
Federal Libraries Committee, 60
Federation for Documentation (FID), 128
Fish and Wildlife Service, of Department of
   Interior, 19
Fisk, James B., 194, 195
Ford Foundation, 125, 157
Forest Service, 25
France
   Biblioteque Nationale of, 2
   National Center for Scientific Research
      (CNRS), 2
French Academy of Sciences, 1
Fry, Bernard M., 43, 62, 82, 158
Fry, George, 73, 103
Fussler, Herman H., 42, 43

Gardin, Jean-Claude, 108
Garfield, Eugene, and Associates, 29
Garwin, Richard L., 206
Geography and Maps, LOC Division of, 15
Geological Society of America (GSA), 123
Geologic Survey, 5, 19
   developments in, 21-23
   establishment of, 10
   expansion of, 30
   information collections of, 10
   joint programs, 121
   satellite earth resources mapping and
      analysis program of, 83
Geo-Ref Service, of AGI, 126
Gilliland, Edwin R., 206
Glazer, Ezra, 100
Goddard Space Center, 65
Gorgas, W. C., 24
Government agencies, responsibilities of,
   203. *See also* Federal agencies
Government Printing Office (GPO), 114, 171
Grants, NSF, 122
Graphic Arts Composing Equipment
   (GRACE), 67, 120
Gray, Dwight E., 158
Green, John C., 34, 36, 99, 159
Greenberger, Martin, 85
Group for Standardization of Information
   Services (GSIS), 70, 94
Guyot, Professor, 19

Hardware, rate of development of, 154
Haskins, Caryl P., 194, 195
Hassler, Ferdinand Rudolph, 20
Hatch Act, 26, 113
Hayden, Ferdinand, 21
HAYSTAQ, 97-99, 175
Hayworth, Leland, 64
Health, Education and Welfare Department,

Index 229

information services of, 81-82
Health Science Resource Center, 141
Heller report, 104-107
Heller, Robert, Associates, 73, 124
Henry, Joseph, 9, 17, 18, 19
Herner and Company, 29, 116-117
Herring, Conyers, 218
Highway Research Board, 178
Highway Research Information Center, 117, 127
Holbrook, George E., 218
Hollerith, Herman, 12
Hookway, Harry, 142
Horning, Donald F., 64, 206
House of Representatives, Select Committee on Government Research, 56. *See also* Congress, U.S.
Hughes legislation, 26
Hughes, Patrick, 135
Humphrey, Hubert H., 55
Hydrographic Office, 23, 24

ICIREPAT. *See* Committee for International Cooperation in Information Retrieval among Examining Patent Offices
Indexing
  Heller report on, 104-105
  of patents, 99
  and technological innovations, 70
Indexing services, 72
  access to, 212
  MEDLARS, 67, 68, 69, 80, 81, 120, 152, 160
  and technological innovations, 47, 66
Indexing terms, 67
Index Medicus, 23, 67, 116, 152, 160, 172
Indians, American, and Smithsonian Institution, 19
Industrial Information Branch, of AEC, 47, 62, 175
Industry
  information, 154
  technical information for, 87
Information activities
  chronology of, 171-179
  cooperative arrangements for, 123
  federal assistance to, 121-125
Information agencies, federal, 5. *See also* Federal agencies
Information Analysis Centers, of DOD, 117
Information centers
  proposed, 202
  sci-tech, 3, 147, 150
Information Policy Group (IPG), of OECD, 144, 159
Information Policy and Program, of DOT, 84
Information Processing Center of NASA, 161
Information sciences, 147
  cooperative support for, 102-103
  funding of, 95
  R&D in, 93-94
Information services
  classical, 213-216
  cooperative arrangements for, 116
  DHEW, 81-82
  Group for Standardization of (GSIS), 70, 94
  growth of, 29
  purpose of, 150
  Soviet, 53, 137-139
Information Services in Physics, Electrotechnology, Computers and Control (INSPEC), 142
Information transfer process, 200
"Inquiry, The," 113
Institute of Medical Sciences (IMS), 125
Institute for Scientific Information, 29
Interagency Group for Research on Information Systems (IGRIS), 94
Interdepartmental Committee for Research and Development, 37
Interior, Department of
  international programs of, 133
  libraries of, 10, 50
International Atomic Energy Agency (IAEA), 48, 134, 143
International Catalogue of Scientific Literature, 20
International Center of Scientific and Technical Information, in Moscow, 139, 143
International Conference on Scientific Information, Soviet participation in, 139
International Council of Scientific Unions (ICSU), 107, 108, 127, 135, 136, 178. *See also* Committee for Data for Science and Technology
International Exchange Service, of Smithsonian, 17-18
International Geophysical Year (IGY), 127, 135, 143

International governmental organizations
    (IGOS), 143-144
International Meteorological Committee
    (IMC), 135
International Meteorological Organization
    (IMO), 135
International Nuclear Information Service
    (INIS), 7, 48, 134
International organizations, and information
    systems, 5
International Standard Book Numbering
    System (ISBN), 134
International Standard Serials Numbering
    System (ISSN), 134, 144
Israeli Center for Scientific Translation, 133

Japan
    Information Center for Science and
        Technology (JICST), 2, 53-54
    National Diet Library, 2
    Science Council of, 1
    Science and Technical Agency, 2
Jefferson, Thomas, 3
    classification scheme of, 14
    library of, 13, 113, 171
Jet Propulsion Laboratory, 65
Jewett, Charles Coffin, 17
Johns Hopkins University
    Applied Physics Laboratory, 29
    Welsh Medical Library of, 49
Joint Commission on Scientific and Technical Communication, 210
Journals
    abstracting, 189
    primary, 188, 190, 192

Katz, Donald L., 218
Katz Report, 140
Kelly, J. Hilary, 57
Kelly, Mervin J., 99
Kennedy, John F., 57
Kennedy Report, 221-224
Key word lists, 89
Killian, James R., Jr., 56, 185, 194, 195
King, Clarence, 21, 22
King, Gilbert W., 70, 99, 118
Kistiakowsky, George B., 194, 195, 206
Knox, William T., 59, 84, 159

Land, Edwin H., 194, 195
Land-Grant Act, 25
Languages

Russian, 188
    of scientific communication, 61
Larkey, Sanford, 49
Lawrence Radiation Laboratory, 115
Legislative Reference Service, 15-16
Librarian of Congress, 15, 131
Libraries. *See also specific libraries*
    DOA, 25-26
    governmental, 5
    national, 2
    regional centers for, 114
    role of, 214
    sci-tech, 4, 147
Library of Congress, 2-3, 120
    collections of, 148
    development of, 12-16
    establishment of, 171
    growth of, 10
    information services of, 148
    joint serials data program, 163
    PL-480 centers of, 133
    postwar developments in, 40-42
    R&D efforts of, 90
    Science and Technology Project, 32, 150
    Smithsonian collection in, 9, 14-15
    study groups of, 118
Library sciences, 147. *See also* Information
    sciences
Licklider, J. C. R., 218
Licklider Report, 177
Linder, Clarence H., 218
Lister Hill National Center for Biomedical
    Communications, 80, 157, 178
Listomatic camera, 70, 152
Lockheed Corporation, 83, 164
Loeb, Robert F., 206
Lovell, Joseph, 23, 171
Lowry, W. Kenneth, 159
Luntz, Jerome D., 219

McCellan, John F., 55
McCoy, H. M., 31
Machine translation, 71, 89, 129
MacLeod, Colin M., 206
Magoun, H. W., 219
Manhattan District Editorial Advisory
    Board (MDEAB), 42
Manhattan District Project, 30, 115, 150
Man-in-Space Center, 65
MARC (Machine Readable Catalog) system,
    70, 134, 153
Marshall, George C., 32

Martell, Charles, 57, 58
Marzke, Oscar T., 219
Mathematical Reviews, 20
Maury, Matthew F., 21, 24
Max Planck Institute, 1
Mechanized Information Processing, 202
Medical Library Assistance Act, 68, 80, 177
Medical Literature Analysis and Retrieval System. *See* MEDLARS
Medical Sciences Information Exchange (MSIE), 37-39, 127, 158, 174
MEDLARS (Medical Literature Analysis and Retrieval System) program, 67, 68, 80, 152, 160
    GRACE component of, 69
    MEDLARS II, 81
    systems development for, 120
"Memex," 97
MESH (Medical Headings Subject File), 81
Meteorological Abstracts, 123, 174
Microcard Corporation, 47
Microcards, 152
Microfiche, 152
Microfilm services, 26, 151, 153
Microforms, 66, 164
    federal adoption of, 119
    set of standards for, 58
Military projects, early, 23-24
Military Sci-Tech Information Services, 50
Miller, Eugene, 159
Mitre Corporation, 29
Mohrhardt, Foster E., 160
"Monthly List of Translations," 37
Morril Act, 25, 113
Multilith reproduction, 94
Museum of History and Technology, of Smithsonian Institution, 18
Museum of Natural History, of Smithsonian Institution, 9, 18

National Academy of Engineering (NAE), 125
National Academy of Science (NAS), 22, 113, 125-130
    cooperative programs with federal agencies, 7
    publications of, 129
    SATCOM Report of, 207-219
National Advisory Committee on Aeronautics (NACA), 5, 83
    expansion of, 30
    research facilities of, 115

National Aeronautics and Space Agency (NASA), 65, 175, 209
    international information exchange program of, 134
    research facilities of, 115
    sci-tech information system of, 86-87
National Agricultural Library, 66, 177
    joint serials data program, 163
    R&D efforts of, 90
National Bureau of Standards (NBS), 3, 114
    HAYSTAQ program of, 97-99
    NSDRS of, 117
    R&D efforts of, 90
National Cancer Institute (NCI), 81-82
National Climatic Data Center, 84
National Committee for Aeronautics, 20
National Committee on Critical Data, British, 142
National defense, impact of modern science and technology on, 200
National Defense Education Act, 54, 63, 118, 185
National Federation of Science Abstracting and Indexing Services (NFSAIS), 73, 123
National Fine Arts Collection, of Smithsonian Institution, 18
National Geophysical Data Center, 84
National Institution for the Promotion of Science, 18, 113
National Library of Medicine, 5, 23, 67, 155, 158, 160
    development of, 10
    expansion of services, 80
    forerunner of, 24
    joint serials data program, 163
    MEDLARS program, 68 (*see also* MEDLARS program)
    systems development for, 120
National Library of Medicine Act, 175
National Marine Fisheries Service, 162
National Medical Audio-Visual Center, 80
"National Nuclear Energy Series," 45
National Oceanographic and Atmospheric Agency (NOAA), 21, 83, 90, 162, 179
National Oceanographic Data Center, 62, 84, 162
National Operational Meteorological System, 135-136
National Referral Center for Science and Technology, of LOC, 63

National Research Council of the National Academy of Sciences (NAS/NRC), 37, 125
National Science Foundation (NSF)
　expansion of, 30
　R&D support of, 95-96
　Science Information Service of, 54
　and systems development, 124
National Science Library, Canadian, 140-141
National Standard Data Reference System (NSDRS), 58, 117, 135, 156, 177
National Systems Task Force, of COSATI, 59, 107, 159
National Technical Information Service (NTIS), 5, 37, 84, 90, 179
　developments in, 33-37
　Xerox equipment used by, 120
National Weather Service, 16, 18, 23, 62
Naval Observatory, 23, 24
Navigation charts, publication of, 20
Navy Department, early projects of, 24
Navy Hydrographic Office, 50
Navy Research Section (NRS), 32, 41
Neville, Leslie E., 31, 32
Newman, James, 187
Newmark, Nathan M., 219
Newton, Issac, 12
National Institutes of Health (NIH), 37, 80, 81. *See also* National Library of Medicine
NLM. *See* National Library of Medicine
Nomenclature, geological, 22
Notices to Mariners, 24
Nuclear science, regional libraries for, 115
"Nuclear Science Abstracts (NSA)," 43, 46, 47

OASIS (Oceanic and Atmospheric Scientific Information System), 179
Office of Critical Tables, of NAS/NRC, 126
Office of Documentation, 130, 176
Office of Management and Budget, 84, 85
Office of Naval Research, 88
　expansion of, 30
　R&D efforts of, 90
Office of Science Information Service (OSIS), NSF, 5, 56, 63-64, 101, 110, 123, 156, 158
　budget of, 74
　interdepartmental translations program of, 132
Office of Science and Technology (OST), 176

　COSI of, 101
　formation of, 64
Office of Scientific Information (OSI), of NSF, 63, 161, 175
Office of Scientific Research and Development (OSRD), 3, 31, 173
Office of Scientific and Technical Information (OSTI), 135, 141-142
　of NASA, 150
　of United Kingdom, 2
Office of Technical Services (OTS), of Commerce Department, 34-35, 62, 150
　cooperative exchange projects of, 132
　legislative base for, 4
Office of Weights and Measures, 3
Offset printing, 94
Operational systems, and new technologies, 85-88
Organization for Economic Cooperation and Development (OECD), 143

Page charges, 122
Pan-American Health Organization, 81
Panofsky, Wolfgang K. H., 206
Patent Act, 11
Patent Office, 5
　agriculture component of, 25
　information collections of, 10
　R&D program of, 96-100
　sci-tech library of, 11-12
Personnel, federal, in sci-tech information, 155-161
Petroleum Production Board, 30
Photocharger, 160
Photoclerk, 152, 160
Photoduplication Service, of LOC, 148
Photographic copies, early use of, 173
Photon, Inc., 120
Pickering, William H., 219
Pierce, Benjamin, 21
Pinchot, Gifford, 25
Powell, John Wesley, 19, 21, 22
President's Science Advisory Committee (PSAC), 54, 55
　Baker panel, 100-101
　Baker Report, 185-195
　role of, 205
　Weinberg Report, 102-103, 199-206
Press, Frank, 206
Private foundations, access to BSIE, 39
"Project Lex," 67, 71, 118

Professional societies, 2, 72, 122. *See also specific societies*
Publications Board, 33-34
Public Health Service, 37
  expansion of, 30
  international programs of, 133
Pucinski, Roman C., 56
Punchcard systems, 66, 152, 153
Purcell, Edward M., 194, 195, 206
Putnam, Herbert, 15

Rabi, Isidor I., 194, 195
Rapid reproduction techniques, 66, 87, 119
RECON system, 83, 87, 134, 153, 163
Recurring Demand Searches, 178
Reed, Walter, 23
Remington-Rand Corporation, 121
Reprographic techniques, 73, 86
Research addressed to National Needs (RANN), 83
Research and Development (R&D)
  growth of, 29
  information sciences, 88, 93-94
  nonfederal support for, 114-118
Research in Education, 82
Research Information Center and Advisory Service on Information Processing (RICASIP), 176
RESPONSA service, 152
Riegel, Byron, 219
Robertson, H. P., 194, 195
Rockefeller Foundation, 125
Rogers, Frank B., 67, 160
Rome Air Development Center, 74, 88
Royalty systems, 104
Russian language, 188

Salisbury, Morse, 44
Sandia Laboratory, conference of AEC librarians at, 46-47
SATCOM (Committee on Scientific and Technical Communications), 129
SATCOM Report, 207-219
Sauter, Hubert, 160
SCAN service, 152
Science Abstracts, 20
Science Division, of LOC, 41
Science Information Exchange, 58, 64
Science Information Service of NSF, 26, 54, 176, 185, 193
Science and Technology Division, of LOC, 5
Science and Technology Project (STP), of LOC, 41, 161
Scientific information, problem of, 187-188
Scientific journals, primary, 188, 190, 192
Scientific societies, 2, 192
Scientists, and federal R&D programs, 94
Sci-tech literature
  increase in, 71
  quantity of published, 60-61
  subsidy of, 122
SEAC computer, 97, 98
Sea Grant Program, 162
Searching techniques, development of, 97
Seitz, Frederick, 206
Senate, U.S. *See also* Congress, U.S.
  Appropriations Committee, 96
  Subcommittee on Reorganization and International Organization, 55
Shannon, Robert, 47
Shaw, Ralph R., 49, 160
Sherrod, John, 63, 160
Smithson, James, 16, 17
Smithsonian Institution
  development in, 16-20
  exchange program of, 5, 131, 147
  international programs of, 133
  publications, 19-20
  scientific library of, 14-15
  sci-tech library of, 9
  working library of, 17
Smithsonian Science Information Exchange, 5, 147
Snow, Ice and Permafrost Establishment (SIPRE) of U.S. Army Corps of Engineers, 41, 149, 160, 174
Snyder, John W., 34
Software, development of, 154, 202
Sorokin, Yuri N., 139
Special Libraries Association (SLA), 36-37
Spencer, R. A., 100
Spofford, Ainsworth R., 13
Sputnik, launching of, 53
Standard Data Reference System, 91. *See also* National Standard Data Reference System
Standardization
  developments in, 154
  for information processing, 163
  problems of, 210
Stanford Research Institute, 29
STAR (*Space Technology and Research*) abstract bulletin, 65
State Service for Standard Reference Data

(GSSSD), of USSR, 135
Stearns, John F., 63, 161
Steere, William C., 219
Stegmaier, Robert B., 161
Step cameras, 66. *See also* Technology
Sternber, George M., 23
STINFO, 221-224
Study groups, nonfederal, 118-119
Subject categories, changes in, 74. *See also* Classification systems
Swanson, Don R., 219
Systems Development Corporation (SDC), 29, 59, 164

Taube, Mortimer, 31, 149, 161
"Tech Briefs," joint program of, 87
Technical Abstract Building (TAB), 153
Technical Information Section/Service, of AEC, 41, 42-48, 62, 157, 174
   exchange agreements of, 133
   function of, 44
Technical News Service, 31
Technology
   adoption of new, 119-121
   computer, 70
   innovations in, 66, 151-153
   and operational advances, 85-88
   step cameras, 66
Technology Utilization Center, of NASA, 115-116
Technology Utilization Program, 87
Telecommunications, 75, 119
Thesauri, 89, 118
Thompson, Alberto F., 42, 44, 45, 63, 161
Thum, E. E., 46
TIROS, 135
Toxicological Information Center, 80, 178
Translating, mechanization of, 71, 89, 129
Tukey, John W., 206, 219
Tumbleson, Robert C., 63
Tuve, Merle A., 219

Unesco, 107, 108-109, 143, 178
Union of American Biological Societies, 126
"Union List of Serials," 189-190
UNISIST, 6, 7, 107-110, 134, 143, 178
United Kingdom
   British Library of, 2
   sci-tech information programs of, 141-143
Univac system, 121, 152
United Nations
   Educational, Scientific, and Cultural Organization (UNESCO), 107, 108-109, 143, 178
   Industrial Development Organization (UNIDO), 143
   and information sciences, 108
USSR
   Academy of Sciences, 1
   GSSSD of, 135
   Lenin Library, 2
   scientific competition of, 53
   sci-tech information programs of, 137-139
   and VINITI, 2, 53, 137
   and world information services, 110

VINITI. *See* All-Union Institute of Scientific and Technical Information

Warren, Stafford, 56, 177
Waterman, Alan T., 54, 57
Watterson, George, 13
Weather Bureau, 5, 123, 162, 172
   expansion of, 30
   information collections of, 10
Weather satellites, introduction of, 135
Weather service, of Smithsonian Institution, 9. *See also* National Weather Service
Weekly Title List, of TIS, 46
Weinberg, Alvin M., 55, 102, 206
Weinberg panel, 57
Weinberg Report, 102, 177, 199-206
Weiss, Paul A., 194, 195, 219
Wenk, Edward, 56, 63
Wheeler, George M., 21
Wiesner, Jerome, 55, 57, 194, 195, 206
Wiley, W. Bradford, 219
Wilkes, Charles, 24
Williams, Van Zandt, 219
Wood Products Research Laboratory, 25, 114
Woodward, J. J., 23
World Data Centers, 127, 129, 135
World Health Organization
   and MEDLARS program, 81
   NCI cooperation with, 82
World Information System, 107
World Meteorological Organization (WMO), 136, 143
World Serials Data Office, 110

World War II, and sci-tech information centers, 31
World Weather Watch, 136, 143, 178
Wright, Irving S., 219
Wysocki, Adam, 107

Xerox Corporation, 119
X organization, 73, 106

Yngve, Victor H., 99
York, Herbert, 194, 195
Young, Yolande D., 43

Z 39 Committee, of American National Standards Institute, 163
Zacharias, Jerrold R., 194, 195, 206
Zwemer, Raymund, 41